ANIMAL SCIENCE, ISSUES AND PROFESSIONS

CARNIVORES

SPECIES, CONSERVATION, AND MANAGEMENT

ANIMAL SCIENCE, ISSUES AND PROFESSIONS

Additional books in this series can be found on Nova's website under the Series tab.

Additional e-books in this series can be found on Nova's website under the e-book tab.

ANIMAL SCIENCE, ISSUES AND PROFESSIONS

CARNIVORES

SPECIES, CONSERVATION, AND MANAGEMENT

FEDERICO I. ÁLVARES
AND
GUILHERME E. MATA
EDITORS

nova publishers
New York

For permission to use material from this book please contact us:
Telephone 631-231-7269; Fax 631-231-8175
Web Site: http://www.novapublishers.com

Library of Congress Cataloging-in-Publication Data

ISBN 978-1-62257-323-3

Published by Nova Science Publishers, Inc. †New York

CONTENTS

PREFACE

In this book, the authors present current research in the conservation, species and management of carnivores. Topics discussed include the adaptation, impact and management of the raccoon dog in Europe; human perceptions of cougars in Canada and El Salvador; the impact on wildlife conservation of emerging protozoal tick-borne diseases of canids; mitigating conflict between humans and large carnivores in carnivore conservation; the management and conservation of wolves in the Iberian Peninsula; and factors affecting small and middle-sized carnivore occurrence and abundance in Mediterranean agricultural landscapes in Southern Portugal.

Chapter 1 - Piroplasmosis (or babesiosis) and hepatozoonosis are the two main protozoal tick-borne diseases occurring in the Canidae. The parasites causing these diseases live in vertebrate blood cells and may induce severe pathologies. Hematozoan infection is widespread in wild and domestic canids. Hence, both piroplasmosis and hepatozoonosis must be taken into account in any future planning for the management of wildlife resources.

The use of microscopy for diagnosis of hematozoans has proven to be an insensitive method. Before the development of molecular techniques for identification of protozoa, the canine piroplasmid species were limited to *B canis canis*, *B. canis vogeli*, *B. canis rossi* and *B. gibsoni.* However, diverse contributions (based on molecular procedures) published in the last years have shown that there is a broader spectrum, which includes *Babesia* (*Theileria*) *annae*, *T. equi*, *T. annulata*, *Theileria* sp, *B. caballi*, *B. conradae*, *Rangelia vitalii* and *Babesia* sp. Among all the aforementioned species, only *Babesia* (*Theileria*) *annae,* members of the *B. canis* complex and some *Theileria* sp have been found in wild canids. In addition, reports on the experimental infections of coyotes with the species *B. gibsoni* suggest that wildlife species may be reservoir for more piroplasmids (hitherto undetected). For hepatozoonosis in wild canids, the situation is quite similar. The disease was thought to be caused only by *H. canis*. However, molecular methods have demonstrated that some new species (*H. americanum*, *H. felis* and *Hepatozoon* sp) are capable of infecting dogs and their kin. Such great advance in epizootiology of hematozoans has been based on standard PCR, Reverse Line Blotting (RLB) and real-time PCR plus DNA sequencing techniques. In this chapter the authors describe the validation of two real-time PCR methods for simultaneous diagnosis of canine tick-borne diseases (caused by piroplasmids and/or *Hepatozoon*) using FRET probes. There are no more than 6 endangered (or critically endangered) canine species. It seems that in general the threat on these species is coming from humans rather than hematozoans. The maintenance of epizootiological surveillance campaigns on wild canids is important, since these animals may serve as sentinel species, indicative of infections threatening domestic dogs (and in some cases humans). Conservation of wild canid species is not always a priority. Instead, population control is ocasionally carried out to limit the expansion of some species (such as the red fox, *Vulpes vulpes*).

Chapter 2 - Raccoon dog (*Nyctereutes procyonoides*) is an alien carnivore in Europe. It is today widespread and common in many parts of the continent. Here the authors review its adaptations to various environmental and climatic conditions, impact on the native fauna and management of the populations. The raccoon dog is very adaptable due to its versatile diet, ability to gather large fat reserves in autumn and to hibernate in areas where winters are harsh. It also has higher than expected reproductive potential, which compensates the high mortality rate of juveniles. Raccoon dogs are monogamous and the male participates in pup rearing. Since no helpers have been observed, juveniles usually disperse during their first autumn and some of them wander far. In a colonizing population also adults have a tendency to wander long distances. Due to massive and dispersed introductions, raccoon dogs are characterized by high genetic variation. Raccoon dogs fitted well into the guild of medium-sized carnivores and share their home ranges with badgers (*Meles meles*) and foxes (*Vulpes vulpes*), and especially raccoon dogs and badgers partly use the same habitats, such as deciduous forests and fields. Raccoon dogs often settle in active badger setts to avoid predation and unfavourable weather conditions. Raccoon dogs may partly overlap their diet with that of native carnivores. Raccoon dogs can spread diseases and parasites, such as rabies and sarcoptic mange, to other carnivores due to their frequent opportunities of contact.

The raccoon dog is a true omnivore and eats anything it can find and catch. Although it sometimes preys on birds, there is no firm evidence that it affects bird populations. The occurrence of birds in its diet shows great spatial and seasonal variation but in most areas is fairly low. Raccoon dogs may have, however, an impact on amphibian populations, especially on islands.

Raccoon dogs are actively hunted in many countries. In Finland, the annual hunting bag is around 160,000 individuals, i.e. about 50% of the autumn population. In spite of many campaigns to reduce the population, the annual hunting bag has steadily increased during the past two decades in Finland. The hunting bag has also increased exponentially in Germany since the early 1990s. Due to their adaptability, large distribution and secretive life, raccoon dogs are very difficult to eradicate after they have established in an area.

Chapter 3 - The relationship between people and the cougar Puma concolor and other larger predators in Canada (brown bears Ursus arctos, black bears Ursus americanus) and Latin America (jaguars Panthera onca) is examined in a comparative mode for cases in Kamloops and Vancouver Island, Canada and El Salvador. Literature derived hypotheses mooting greater sympathy for carnivores from women and young people, greater fear from women and older people, and greater fear of the larger brown bears and jaguars over cougars are tested. The results show that the hypotheses are both supported and refuted. The size of the predator is as important as common ideas about its behaviour, and troublesome invasive behaviour (mostly from bears) is at least as important as predatory behaviour (largely from cougars and jaguars). However, consistent with hypotheses, women and older people were in many instances more afraid of cougars and other carnivores, and older people were generally less tolerant of carnivores. Differences included opinions on positive impacts on quality of life and useful functions of cougars (favoured by men in El Salvador, both men and women in Vancouver Island, and women in Kamloops), and danger to adults, children and other animals (favoured by women more than men in El Salvador for all groups, women more than men in Vancouver Island for children and animals, not adults, and women more than men in Kamloops for adults and children, not animals). For age, older people in El Salvador were generally less tolerant than younger people of large carnivores, similar to Vancouver Island, but not Kamloops, where there were no age-related opinions of the danger from carnivores towards people. The cougar's elusive behaviour limited comparisons with bears and jaguars. Awareness of such issues may assist wildlife conservation.

Chapter 4 - Mitigating conflict between humans and large carnivores is one of the most pressing and intractable concerns in conservation. Yet, there has been surprisingly little effort devoted to incorporating risk assessments of

conflict in carnivore conservation and land-use planning. Because human-carnivore conflict can have far-reaching societal and environmental impacts, attention to the 'conflict–conservation nexus' should become integrated into national and global environmental policy-making. However, *how* 'the nexus' is defined, elucidated, and ultimately utilized to forecast and mitigate conflict remains under-explored. Here, the authors discuss the limitations of current knowledge and methodologies available to forecast human–carnivore conflict and suggest a novel heuristic framework that integrates ecological and sociological data to better predict and mitigate conflict, and optimize conservation planning.
Here the authors illustrate the utility of their approach using a case study of carnivore connectivity planning in the southwestern United States. Their approach holds promise as an effective tool for use in carnivore conservation by allowing decision-makers to prioritize planning efforts by integrating biological suitability, threat of conflict, and societal acceptance.

Chapter 5 - The gray wolf (Canis lupus) is a generalist species whose distribution was originally the widest among wild carnivores but suffered a marked regression due to human persecution during the 19th and 20th centuries. Legal protection after 1970 allowed wolves to recover significant parts of their range, including, in some cases, heavily humanized and modified landscapes. Nowadays, many wolf populations keep expanding, although the viability of many others is still at risk due to small population sizes and loss of genetic diversity. The Iberian Peninsula holds Western Europe's largest wolf population, which is currently divided into two distinct nuclei: a large and practically continuous one spreading through most of the northern half of this region, and a smaller and subdivided one in a southern mountainous area. A distribution model based on the whole Iberian wolf distribution overlooked the southern nucleus, suggesting biogeographic differences between them. This can happen due to local adaptation events, and points to the need for caution when modelling species or populations with disjunct distributions. When the authors modelled both wolf nuclei separately, the southern nucleus showed indeed different environmental and biogeographical properties, with a trend towards a metapopulational structure. Here the authors discuss the implications of these models for the conservation and management of wolves in the Iberian Peninsula and of other species facing similar biogeographic situations elsewhere.

Chapter 6 - Carnivores have a key role in ecosystems and their populations are declining at an increasing rate. Habitat loss and degradation through agricultural practices are among the most serious menaces affecting carnivore survival. Agricultural landscapes are dominant in Western Europe and in several places worldwide are changing quickly into more intensive practices. Therefore, it is important to understand how small carnivore occurrence and abundance is affected by these changes in order to find ways to manage agricultural and grazing systems in a sustainable way, allowing both biodiversity and production to co-exist. Here the authors present three case studies in Natura 2000 areas, mainly covered by private agricultural land (livestock, cereal crops, oak, pine and eucalyptus plantations) in southern Portugal.

In areas of dominant traditional agro–silvo–pastoral systems (montado) small and middle-sized carnivores tend to occur in a mosaic of montado and shrubs. The authors found that the occurrence of one of the forest species was positively related with the density of trees and shrubs, soil organic matter content, and Shannon's index of vegetation vertical diversity. The presence of livestock and extension of game-estate areas also seem to influence carnivore occurrence.

In areas of extensive cereal crops, the presence of shrubs and age of forest plantations play an important positive role for the carnivore community. Mainly the author's results suggest that maintaining a sustainable mosaic embracing montado, shrubland and open land areas, may allow higher species richness and abundance by enhancing connectivity between crucial areas. The implementation of this kind of agricultural practices considering the landowners needs is the key issue to achieve the main carnivore conservation goals in Southern Portugal.

In: Carnivores
Editors: F. I. Álvares and G. E. Mata
ISBN 978-1-62257-323-3

Chapter 1

EMERGING PROTOZOAL TICK-BORNE DISEASES OF CANIDS (PIROPLASMOSIS AND HEPATOZOONOSIS): IMPACT ON WILDLIFE CONSERVATION

Angel Criado-Fornelio
Parasitology Laboratory, Microbiology and Parasitology Department, Faculty of Pharmacy, Universidad de Alcalá, Madrid, Spain

ABSTRACT

Piroplasmosis (or babesiosis) and hepatozoonosis are the two main protozoal tick-borne diseases occurring in the Canidae. The parasites causing these diseases live in vertebrate blood cells and may induce severe pathologies. Hematozoan infection is widespread in wild and domestic canids. Hence, both piroplasmosis and hepatozoonosis must be taken into account in any future planning for the management of wildlife resources.

The use of microscopy for diagnosis of hematozoans has proven to be an insensitive method. Before the development of molecular techniques for identification of protozoa, the canine piroplasmid species were limited to *B canis canis*, *B. canis vogeli*, *B. canis rossi* and *B. gibsoni*. However, diverse contributions (based on molecular procedures) published in the last years have shown that there is a broader spectrum, which includes *Babesia* (*Theileria*) *annae*, *T. equi*, *T. annulata*, *Theileria* sp, *B. caballi*, *B. conradae*, *Rangelia vitalii* and *Babesia* sp. Among all the aforementioned species, only *Babesia* (*Theileria*) *annae,* members of the *B. canis* complex and some *Theileria* sp have been found in wild canids. In addition, reports on the experimental infections of coyotes with the species *B. gibsoni* suggest that wildlife species may be reservoir for more piroplasmids (hitherto undetected). For hepatozoonosis in wild canids, the situation is quite similar. The disease was thought to be caused only by *H. canis*. However, molecular methods have demonstrated that some new species (*H. americanum*, *H. felis* and *Hepatozoon* sp) are capable of infecting dogs and their kin. Such great advance in epizootiology of hematozoans has been based on standard PCR, Reverse Line Blotting (RLB) and real-time PCR plus DNA sequencing techniques. In this chapter we describe the validation of two real-time PCR methods for simultaneous diagnosis of canine tick-borne diseases (caused by piroplasmids and/or *Hepatozoon*) using FRET probes.

There are no more than 6 endangered (or critically endangered) canine species. It seems that in general the threat on these species is coming from humans rather than hematozoans. The maintenance of epizootiological surveillance campaigns on wild canids is important, since these animals may serve as sentinel species, indicative of infections threatening domestic dogs (and in some cases humans). Conservation of wild canid species is not always a priority. Instead, population control is ocasionally carried out to limit the expansion of some species (such as the red fox, *Vulpes vulpes*).

1. Introduction

Usually pathogenic infections of wildlife are situated in a negative context, as a cause of wildlife endangerment and possibly extinction. The hematozoan parasites of canids may cause sporadic episodes of mortality, but usually this occurs mostly in young animals or alternatively in individuals under stress conditions or infected with other pathogens (Harrus and Baneth, 2005; Penzhorn, 2006). However, it is important to underline that from a biological point of view, parasites are not positive or negative; they just intervene as part of the natural selection process (Anderson and May, 1978). Moreover, usually parasites are not prone to kill the host, since this would eventually lead to their own extinction. This is more evident in pathogens showing high parasitic specificity (Dunn *et al.*, 2009), a feature not always evident for the main hematic protozoa found in wild canids in Europe (*B. annae* and *Hepatozoon* spp). Considering these facts, the present chapter may seem as superfluous. However, it must not be forgotten that the growing anthropization of most natural habitats drives wildlife species near to both humans and domestic animals (Penzhorn, 2006). Therefore, wild canids and dog (or other domestic mammals) may share pathogens in certain in geographic locations. Indeed, wild canids rarely visit urban areas, with the possible exception of the red fox, which is an opportunistic predator and scavenger. The visits of foxes (or other canids like coyotes) to human settlements might be increasing everywhere due to global warming or local drought periods (Criado-Fornelio *et al.*, 2000). But anyhow, parasite transmission may also occur in periurban or even rural regions. In consequence, it could be possible the exchange of parasites of dogs and foxes in some particular areas (Gabrielli *et al.*, 2010). Only in very few cases the microbial pathogens of domestic dog have been transmitted to wild canids (Murray *et al.*, 1999). Likewise, hematozoans of carnivores very seldom affect humans (Shulkina *et al.*, 2004). Although the transmission risk exists, it is extremely difficult to evaluate.

The tick-borne hematozoans of wild canids and dogs belong to four apicomplexan genera: three genera (*Babesia*, *Theileria*, *Rangelia*) in the order Piroplasmida and the genus (*Hepatozoon*) in the Suborder Adeleorina. These two groups (piroplasmids and hepatozoons) differ in their way of transmission. Piroplasmids are injected by ticks in vertebrates, while *Hepatozoon* spp reach their hosts by tick ingestion and also by eating infected prey or by vertical transmission (Baneth 2011, Solano-Gallego and Baneth, 2011). Pathogenic effects of these parasites have frequently been recorded in dogs, in contrast with wild canids where the harmful effect of hematozoans is hard to detect.

Drugs for the treatment of hematozoan-infected dogs are available, but such kind of sanitary intervention is rarely performed on wild animals. Likewise, there are some vaccines for dog piroplasmids (e.g. *B. canis*), but they are not intended for wild canids (Schetters *et al.*, 2009). With regard to the endangered wild canids, six species are especially vulnerable (IUCN, 2011): *Canis rufus* (red wolf), *Canis simensis* (Ethiopian wolf), *Cuon alpinus* (dhole), *Lycaon pictus* (African wild dog), *Pseudalopex fulvipes* (Darwin's fox) and *Urocyon litoralis* (island fox). The presence of hematozoans has seldom been reported in these species, but they are present indeed in other non-endangered canid species that may share habitats with them. It is evident that although protozoa may not be a first magnitude threat for wild canids, they should not be totally disregarded (Giannitti *et al.*, 2012; Clancey *et al.*, 2010).

In the last years, several reviews on hematozoans of dogs have been published (Ewing and Panciera, 2003; Irwin, 2009; Baneth, 2011; Solano-Gallego and Baneth, 2011; O'Dwyer, 2011).

Table 1. Main piroplasmid and *Hepatozoon* species diagnosed in wild canids

Hematozoan Species	Morphology	Host(s)	Publication(s)
B. canis rossi	Large piroplasmid, no tetrads or Maltese cross	Jackal (*Canis mesomelas*) African wild dog (*Lycaon pictus*)	Nutall (1912) Matjilla *et al.*, (2008)
B. annae	Small piroplasmid, no tetrads or Maltese cross	Red fox	Goethert and Telford, (2003): Criado-Fornelio *et al.*, (2003b)
Babesia canis canis	Sequence data only – Not yet observed in wild canids	Red fox	Torina *et al.*, (2007)
H. canis	Gamonts similar to those found in dogs	Red fox South American foxes *Lycalopex gymnocercus* and *Dusicyon thous* African wild dog Conceiçao-Silva *et al.*, (1988) and	Criado-Fornelio *et al.*, (2003b) Criado-Fornelio *et al.*, (2006) Matjilla *et al.*, (2008)
Hepatozoon sp BV-2	Sequence data only – Not yet observed in wild canids	Red fox	Dezdek *et al.*, (2010)
H. americanum	Gamonts similar to those found in dogs	Coyote Grey fox	Kocan *et al.*, (1999) and (2000) Allen *et al.*, (2011)
Hepatozoon sp	Sequence data only – Not yet observed in wild canids	*Dusicyon thous* Bush dog, maned wolf	Criado-Fornelio *et al.*, (2006) André *et al.*, (2010)
H. felis	*H. felis* gamonts are slightly smaller than *H. canis*	Pampas grey fox (*Lycalopex gymnocercus*)	Giannitti *et al.*, (2012)
*B. gibsoni** *Experimental infection*	Small piroplasmid, similar to the original dog isolate	Coyote (*Canis latrans*)	Evers *et al.*, (2003)

Emphasis is given to recent publications based on molecular methods of identification (DNA sequencing).

However, far less attention has been paid to hematic protozoa of wild canids. By this reason, these pathogens shall be considered in this chapter. Both existing and emerging hematozoans shall be briefly described, with mention of their morphology, biology, pathologic features, phylogeny, epizootiology, and control issues (with special reference to conservation).

2. Canine Piroplasmids: Morphology and Life Cycle

Piroplasmid of dogs and wild canids have been morphologically classified as small (length < 1.5 μm) or large (length > 1.5 μm) species based on their aspect when the parasites are observed inside host erythrocytes. Canine piroplasmids do not form Maltese cross inside red blood cells, although tetrads have been described for *B. gibsoni* (Namikawa *et al.*, 1988) and *B. conradae* (Kjemtrup *et al.*, 2006). An overview of piroplasmid species found in canids is shown in Table 1. As stated above, three genera have been described in canids (*Babesia*, *Theileria* and *Rangelia*). In the case of theilerids, most species were only detected in dogs by molecular methods, so up to now there is no morphological description available. In addition *Theileria annae* of dogs and foxes (Zhaler *et al.*, 2000, Criado-Fornelio *et al.*, 2003b) has been denominated *Babesia microti*-like or *Babesia* (*Theileria*) *annae* due to its lack of lymphocytic stages (Goethert and Telford, 2003). Criado-Fornelio (2012a) has suggested (in agreement with Clancey *et al.*, 2010) that the name *Babesia annae* is the best option to avoid confusion with some rodent isolates from Japan, which have likewise been identified as *Babesia microti*-like. *Rangelia vitalii* was considered a synonym of *B. canis*, but DNA sequence data clearly showed that it is a valid species (Loretti *et al.*, 2005; Soares *et al.*, 2011).

The life cycle of canine piroplasmids (Figure 1) involves both vertebrates and invertebrates. Dogs or other canids serve as intermediate hosts while Ixodid ticks are definitive hosts (Schmidt and Roberts, 1981). During a blood meal, a *Babesia*-infected tick introduces the parasite sporozoites into the canid host. These inoculated protozoa enter either in leukocytes (*Theileria*, *Rangelia*) or in erythrocytes (*Babesia*) and undergo asexual reproduction. In the case of *Theileria* sp, *T. equi, T. annulata* or *B. caballi*, there is no idea of how the life cycle develops in dogs (Criado-Fornelio *et al.*, 2003b and 2007a; Matjila *et al.*, 2008a; Beck *et al.*, 2009; Fritz, 2010; Dezdek *et al.*, 2010).

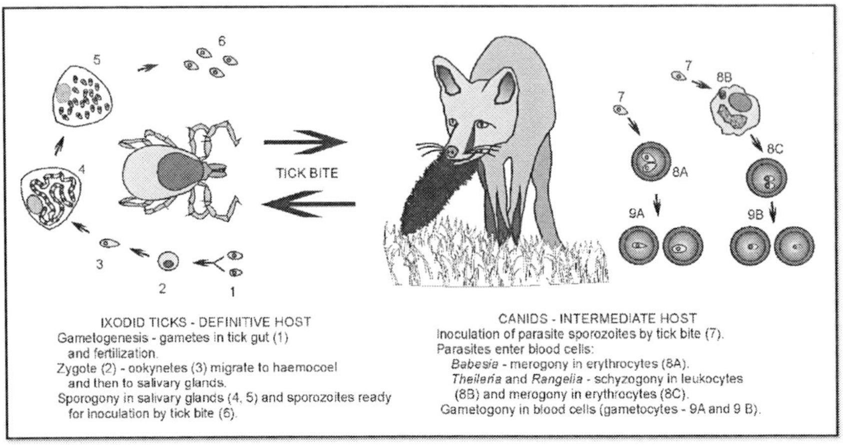

Figure 1. Schematic representation of the life cycle of the canine piroplasmids.

For *Rangelia vitalii*, it seems that there is a stage in leukocytes and endothelial cells of dogs previous to the intraerythrocytic stage (Da Silva et *al.*, 2011). After some cycles of merogony in vertebrate blood cells, piroplasmids differentiate into gametocytes. These are are ingested by a tick, where gametes develop in the gut cells and then fuse to form a zygote. The diploid cell undergoes a multiple fission process in tick gut cells (sporogony). The sporozoites migrate (through the coelom) to the tick salivary glands, where eventually protozoa can restart the life cycle, once a new bite occurs (Kumar *et al.*, 2008). Both trans-stadial and trans-ovarian transmission has been found in *Babesia* species. In contrast, trans-ovarian transmission has not been reported for any *Theileria* species. It seems that all those biological features allow the parasite to remain inside ticks for a high number of generations, which favours parasite survival (Solano-Gallego and Baneth, 2011).

3. Canine Hepatozoons: Morphology and Life Cycle

The morphology and biology of hepatozoons is better known in canids than in felids. There are at least four main species of *Hepatozoon* capable of infecting canis: *H. canis*, *H. americanum*, *Hepatozoon* sp (related to rodent/lagomorph hepatozoons), *H. felis* and diverse *Hepatozoon* species of uncertain affiliation.

The gamont of *Hepatozoon* sp is the diagnostic stage in dog blood (Baneth *et al.*, 1998a). Meronts (schizonts) of *Hepatozoon* sp have been observed in diverse tissues of infected dogs (Baneth *et al.*, 2007). These authors provided a complete description of the morphology of *H. canis* in dogs and in the vector tick (*Rhipicephalus sanguineus*). Likewise, studies carried out in th USA have shown the morphology of *H. americanum* in dogs (Panciera *et al.*, 2001). It seems that *H. canis* lives inside neutrophils while *H. americanum* prefers monocytes. In addition, the latter species undergoes merogony in skeletal muscle, whereas the former may parasitize diverse sites (such as hemolymphatic tissues and visceral organs).

The meront stages of both species may also be recognized, since *H. americanum* shows "onion skin" cysts, whereas the morphology of *H. canis* merozoites in the meront resembles a "wheel spoke" (Ewing and Panciera, 2003). However, little work on parasite morphology in host species other than dog has been published up to now.

H. felis has been found in a South American fox (*Lycalopex gymnocercus*) and the protozoal muscle cysts were described as similar to those of *H. americanum*, albeit slightly smaller in size (Giannitti *et al.*, 2012). Vojta *et al.*, (2009) pointed out that *H. felis* has been found in dogs, but such affirmation is inexact, as the Genbank® entry mentioned in the discussion section of their work (EF188809, introduced in Genbank by East *et al.*, 2008) corresponds to a spotted hyaena (*Crocuta crocuta*). This species belongs to the order Feliformia (Eizirik *et al.*, 2010). Although the spotted hyaena may superficially resemble a canid species, indeed it is much more closely related to felids.

Some *Hepatozoon* species have been detected by molecular methods in wild canids (Criado-Fornelio *et al.*, 2006; Matjila *et al.*, 2008b; Dezdek *et al.*, 2010; Allen *et al.*, 2011). However, no data on the morphology or biology of these hematozoans were provided by these authors.

The life cycle of *H. canis* and *H. americanum* has been studied in detail by Baneth *et al.*, (2007) and Ewing *et al.*, (2002), respectively. Canids are the intermediate hosts and ticks (*Rhipicephalus* and *Amblyomma*) are the definitive hosts, although other genera of ticks might be also vectors. For other hepatozoons the definitive host is not known, although Criado-Fornelio (2012b) has suggested that *H. felis* might be transmitted by *R. sanguineus* as well.

Sexual reproduction (final gamete formation plus fertilization with subsequent schyzogony) is linked to the arthropod host, whereas schizogony (merogony) and initial gametogenesis is linked to the mammalian host. Based on previously published data (Vincent-Johnson *et al.*, 1997; Ewing *et al.*,

2002; Baneth *et al.*, 2007) a draft of the life cycle of *H. canis* in dogs is shown in Figure 2. In summary, the life cycle involves canids (as intermediate hosts) and diverse ticks. The brown dog tick, *Rhipicephalus sanguineus* is the definitive host of *H. canis* whereas *Amblyomma maculatum* is the main definitive host of *H. americanum*. The existence of alternative vector ticks should not be totally disregarded (please see the section "Epizootiology" in this chapter).

There have been reports of diverse *Hepatozoon* sp infections in canids, but there is no information available on their possible definitive hosts; indeed ticks, fleas or other arthropods are good candidates (Criado-Fornelio *et al.*, 2006; Vojta *et al.*, 2009).

The life-cycle of hepatozoons is completed by ingestion of an infected tick by the definitive canid host. The ingestion of ticks may occur in several ways, such as grooming, licking, scavenging or by predation. But there are two alternative ways of transmission of *H. canis* both in dogs and wild canids. It seems that some *Hepatozoon* stages living in host tissues (cystozoites) may survive after ingestion and subsequently boost an infection in the new host, thus helping to spread the disease (McCully *et al.*, 1975; Smith, 1996; Baneth and Shkap, 2003). In addition, the existence of vertical transmission of *H. canis* has been demonstrated by Murata *et al.*, (1993).

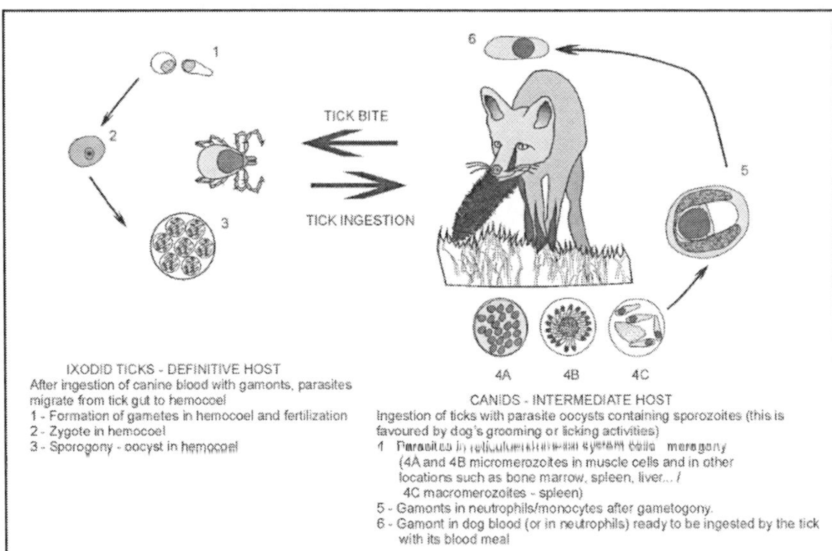

Figure 2. Schematic representation of the life cycle of the canine hepatozoons.

4. Pathology Caused by Piroplasmids and Hepatozoons in Canids

The pathogenic effects of hematozoan infection have been investigated to some extent in domestic canids, although some authors suggest that this important area of research has been poorly addressed up to date (Murray *et al.*, 1999 and Day, 2011).

For piroplasmids, data on the pathology observed in wild canids are scarce, whereas there is much more information is available on the disease in dogs. The presentation of piroplasmosis in dogs depends on the infecting species.

In general, the following findings may be common in symptomatic dogs: haemolytic anaemia, thrombocytopenia, hypoxia, hypertensive shock and multiple organ failure (Solano-Gallego and Baneth, 2011; Da Silva *et al.*, 2011). In addition, blood biochemistry parameters may be significantly altered: azotemia, high serum concentrations of urea and creatinin, together with hyperphosphatemia, hypoalbuminemia, hypercholesterolemia, proteinuria and increased protein/creatinin ratio.

Microscopic analysis of urine may show the presence of hyaline and granular casts (García, 2006). Indeed all these problems may lead to shock and subsequent death of infected animals. With regard to wild canids, most of the information available on piroplasmid infections is related to foxes. Thus, several authors (Criado-Fornelio *et al.*, 2003b, Criado-Fornelio *et al.*, 2007a and Giménez *et al.*, 2009) reported moderate levels of infection caused by *Babesia* (*Theileria*) *annae* in red fox (*Vulpes vulpes*) in Spain. However, most animals were in good condition after capture, with no evident disease signs. Likewise, Birkenheuer *et al.*, (2010) diagnosed moderate prevalence of *Babesia* (*Theileria*) *annae* in foxes from the USA, with no sign of disease. This fact led the authors to conclude that the parasite is not highly virulent for foxes. Interestingly, the coyotes studied in the same area showed no trace of piroplasmid infection.

In Poland, Karbowiak *et al.* (2010) found *Babesia* (*Theileria*) *annae* in just one red fox (0.7%). Likewise, *B. annae* showed low prevalence in red foxes in Croatia (5%) (Dezdek *et al.*, 2010). No detail on the condition of the animals studied was presented in the aforementioned studies in Poland or Croatia.

In Canada, Clancey *et al.*, (2010) reported the first case of piroplasmosis [caused by *Babesia* (*Theileria*) *annae*] in a red fox. Clinical findings consisted in weakness and pale mucous membranes. Hematological findings showed moderate anaemia, low hemoglobin and hematocrit. Blood biochemistry was altered, with increased levels of creatinine, transaminases (ALT, AST) and creatine kinase (CK). The animal was euthanized due to poor prognosis. However, the authors were unsure whether these findings were just due to piroplasmosis or not. It might have been another undetected infection which could have caused the signs observed in the fox. André *et al.*, (2011) reported the serologic detection of *Babesia* spp infection in crab-eating foxes and bush dogs in Brazil, albeit the animals studied seemed to be healthy. In South Africa, a captive wild dog died of acute babesiosis caused by *B. canis rossi* (Colly and Nesby, 1992). However, African wild dogs generally tolerate well this babesial infection (Penzhorn, 2011).

The pathogenic effects of *H. canis* infection have been extensively studied in domestic canids (Pau *et al.*, 1982; Gaunt *et al.*, 1983; Baneth *et al.*, 1995). There are cases where dogs show no sign of disease, but sometimes there are serious complications.

Animals appear lethargic, with fever and extreme loss weight. Hepatoozonosis may be fatal in some cases. There are characteristic hematological changes, such as anemia and leukemia (the latter usually appears only when parasitemia is high). Blood biochemistry may also show alterations, with hyperproteinemia, increased CK (Creatine kinase) and ALP (Alkalin Phosphatase).

Analysis post-mortem of infected dogs usually reveals alterations in liver, lungs and kidney. The occurrence of lesions in periostium is rare in *H. canis*, in contrast with *H. americanum* (Baneth, 2011). The finding of disease signs in wild canids infected by *Hepatozoon* sp is uncommon. Thus Conceiçao-Silva *et al.*, (1988), Criado-Fornelio *et al.*, (2000), Giménez *et al.*, 2009, Karbowiak *et al.*, (2010), Drezdek *et al.*, (2010) and Gabrielli *et al.*, (2010) did not report any disease signs in *Hepatozoon*-infected red foxes in Europe.

Likewise in the New World, Mercer *et al.*, (1988), Criado-Fornelio *et al.*, (2006), André *et al.*, (2010) and Allen *et al.*, (2011) indicated that the coyotes, gray foxes, red foxes or South American foxes they studied presented a good physical condition. However, infections caused by *H. canis, H. americanum* or *Hepatozoon* sp were readily detected in some of the studied wild canids.

In contrast, Giannitti *et al.*, (2012) found clinical signs in a South American fox infected with *H. felis*. Interestingly, the sick animal was also infected by canine distemper virus; therefore disease signs may have been exacerbated by that concomitant infection.

5. Advances in the Diagnosis of Piroplasmids and Hepatozoons in Canids

5.1. Microscopy

Microscopic examination of blood smears has been the classical procedure for identification of canine tick-borne hematozoans (both piroplasmids and hepatozoons). Mercer and Craig (1988) compared three stains (Diff-Quik stain, Giemsa stain and a modification of the naphthol-ASD-chloroacetate esterase stain) and determined that the latter produced the best results in the diagnosis of canine hepatozoons. For piroplasmids, the standard Giemsa stain is preferred in most laboratories. Examination of buffy coat is more sensitive and is considered the routine direct technique in most veterinary clinic laboratories (Otranto *et al.*, 2010). The QBC (Quantitative Buffy Coat) technique is superior to microscopic examination of blood smears (or examination of buffy coat) for diagnosis of hematic protozoa. This method has been extensively employed in diagnosis of human malaria (Gray et al., 1991; Hänscheid, 1999). However, it has been rarely used in veterinary diagnosis, probably due to economic reasons.

5.2. Immunological Procedures

Diverse immunodiagnostic methods for canine babesiosis have been published (Table 2). Serological diagnosis is generally based on the immunofluorescent antibody test (IFAT) or enzyme-linked immuno sorbent assay (ELISA) techniques. Western blot has been used as well, but with much lesser frequence (Baneth 2011; Solano-Gallego and Baneth, 2011).

Historically, the first immunological tests were based on soluble antigens obtained by parasite culture. In the last years, biotechnology has allowed the design of more specific tests by employing methodologies such as recombinant proteins and phage display cDNA libraries (Georgieva and Konthur, 2011). Indeed, engineered antibodies and their fragments are invaluable tools for a vast range of biotechnological and pharmaceutical applications. However, they are facing increasing competition from a new generation of protein display scaffolds, specifically selected for binding virtually any target (Hosse *et al.*, 2006).

Table 2. Review of immunological diagnostic procedures available for tick-borne protozoa of canids

Technique(s)	Target antigen (Protein – mentioned only for RCA)	Parasite detected / Host	Publication
IFAT	PAR	*Babesia canis* / dog	Du Plessis *et al.*, (1990)
IFAT + WB	PAR	*B. gibsoni* / dog	Adachi *et al.*, (1992)
IFAT	PAR	*B. gibsoni* / dog	Yamane *et al.*, (1993)
IFAT + WB	PAR	*B. canis* / dog	Wlosniewski *et al.*, (1997)
WB	PAR	*B. canis* and *B. gibsoni*/ dog	Baneth *et al.*, (1998a)
IFAT	PAR	*Hepatozoon canis* / dog	Baneth *et al.*, (1998b)
IFAT	PAR	*B. canis* and *B. gibsoni*/ dog	Suksawat *et al.*, (2001)
ELISA	PAR	*H. americanum* / dog	Mathew *et al*, (2001)
ELISA	RCA (P29)	*B. gibsoni* / dog	Fukumoto *et al.*, (2003)
ELISA	RCA (P50)	*B. gibsoni* / dog	Fukumoto *et al.*, (2004)
ELISA	RCA (P50 truncated protein)	*B. gibsoni* / dog	Verdida *et al.*, (2004)
ELISA	PAR	*H. canis* / dog	Gonen *et al.*, (2004)
ELISA	PAR	*H. canis* / fox	Fishman *et al.*, (2004)
AD-IC	RCA (P50)	*B. gibsoni* / dog	Verdida *et al.*, (2005)
WB	RCA (P18)	*B. gibsoni* / dog	Zhou *et al.*, (2006a)
WB + ELISA	RCA (AMA-1)	*B. gibsoni* / dog	Zhou *et al.*, (2006b)
IFAT	PAR	*Babesia canis vogeli* / dog	Trapp *et al.*, (2006)
ELISA	RCA (SA-1)	*B. gibsoni* / dog	Jia *et al.*, (2006)
ELISA	RCA (P50)	*B. gibsoni* / dog	Miyama *et al.*, (2006)
ELISA	RCA (P57)	*B. gibsoni* / dog	Aboge *et al.*, (2007)
ELISA	RCA (RAP-1)	*B. gibsoni* / dog	Zhou *et al.*, (2007)
AD-IC	RCA (SA1)	*B. gibsoni* / dog	Jia *et al.*, (2007)
ELISA	RCA (IRA)	*B. gibsoni* / dog	Jia *et al.*, (2008)
ELISA	RCA (TRAP)	*B. gibsoni* / dog	Goo *et al.*, (2008)
ELISA	RCA (TRAP)	*B. gibsoni* / dog	Goo *et al.*, (2009a)
ELISA + WB + IFAT	RCA (P12)	*B. gibsoni* / dog	Goo *et al.*, (2009b)
ELISA	RCA (SA-3)	*B. gibsoni* / dog	Jia *et al.*, (2009)
ELISA	RCA (RAP-1S)	*B. gibsoni* / dog	Terkawi *et al.*, (2009)

Technique(s)	Target antigen (Protein – mentioned only for RCA)	Parasite detected / Host	Publication
AD-BS + PDA	RCA (P50) (Labelled phage display antibodies targeted this protein in blood smears)	*B. gibsoni* / dog	Hirose *et al.*, (2009)
IFAT	PAR	*B. canis* / dog	Furuta *et al.*, (2009)
ELISA	RCA (TRAPf)	*B. gibsoni* / dog	Narantsatsral *et al.*, (2011)

Most tests are directed to detection of antibodies against *Babesia gibsoni*. All the assays mentioned here are only experimental. Abbreviations:

Techniques: IFAT = Immunofluorescent Antibody Technique, ELISA = Enzyme-Linked Immunosorbent Assay, WB = Western Blot, AD-IC = Antigen detection by Immunochromatographic assay - dipstick test, AD-BS + PDA = Antigen detection in blood smears using phage derived antibodies.

Target antigen(s): PAR = soluble preparation of parasites, RCA = Recombinant antigen.

Some commercial IFAT kits (Megacor®, Austria) are available for diagnosis of *B. canis* and *B. gibsoni*. Serum antibodies against *Theileria equi* (which has been found with relative low frequency in dogs from Southern Europe) may be detected with kits based on IFAT (Gentaur®, Belgium) or on ELISA techniques (Eurovet®, The Netherlands). These tests were designed for detection in equines but it should be possible to adapt them for detection in canids.

The development of diagnostic tests for canine hepatozoons is still in an embrionary stage. Attempts to develop IFAT or ELISA tests for diagnosis of canine hepatozoonosis have been published, albeit all of them are still experimental. Baneth *et al.*, (1998b) tested the IFAT technique for diagnosis of *H. canis* in dogs. Matthew *et al.*, (2001) developed an ELISA test for the detection of *H. americanum* in dogs. Both Gonen *et al.*, (2004) and Fishman *et al.*, (2004) employed an ELISA test for diagnosis of *H. canis* in dogs and red foxes, respectively. At this moment there is no commercial immunological test available to diagnose hepatozoonosis in canids.

The use of antigen detection tests has caused a revolution in the diagnosis of human diseases. Unfortunately there had been not enough efforts in the development of antigen detection tests for parasitic protozoa of veterinary interest, with the only exception of *T. cruzi* (Cardinal *et al.*, 2006). Both Verdida *et al.*, (2005) and Jia *et al.*, (2007) reported the development of experimental immunochromatographic antigen detection tests for *B. gibsoni*.

Hirose *et al.*, (2009) published an interesting attempt to develop phage display single chain antibodies for detection of the *B. gibsoni* P50 antigen in blood smears. The results were encouraging, and this contribution may be a significant step towards the development of a commercial immunochromatographic antigen detection test. Dipstick tests are desirable for an easy and quick detection of both *Babesia* and *Hepatozoon* infections (Wong *et al.*, 2011). Unfortunately, no antigen test destined to detect piroplasmids or *Hepatozoon* spp has reached the commercial stage up to date.

In human medicine, a new immunoenzymatic technique denominated as multiple antigen blot assay (MABA) has been employed for simultaneous detection of various pathogens (Noya and Alarcón, 1998).

The technique is a dot blot asay where several antigens (corresponding to different pathogens) are bound to a membrane and then exposed to the test serum. The primary antibodies bound to the membrane are revealed with the help of a secondary (labelled) antibody. If MABA tests were applied for simultaneous detection of infections caused by tick-borne pathogens of veterinary interest, it would be a fascinating diagnostic tool.

The usefulness of immunological tests relies on the hematozoan parasitic fauna present in a certain region. If the number of piroplasmid/hepatozoon species is limited (this may be the case of foxes in Southern Europe) these tests may be telling. When there are several hematozoan species in the area under study, cross-reactivity may become a problem (Solano-Gallego and Baneth, 2011). As indicated above, the development of suitable antigen tests may overcome such problems.

5.3. Molecular Methods

Diverse molecular methods have been developed for the detection of piroplasmids and hepatozoons in the last few years. Before summarizing the published procedures, it is important to underline that Li *et al.,* (2008) found that molecular diagnostic sensitivity was increased by 300% when blood samples were maintained at 37 ºC for three days. The reason for this effect is unclear, but it is a wise idea for those willing to diagnose *Hepatozoon* sp in canids to use such incubation conditions for their blood samples before proceeding with DNA extraction. It is unknown if the incubation step is advantageous for diagnosing piroplasmids.

The available diagnostic assays for tick-borne protozoa of canids are based on the following methods:

1) Standard PCR, Nested PCR, PCR-RFLP or Multiplex PCR with Detection of Amplified Products by DNA Electrophoresis in Agarose Gels

Standard PCR methods for diagnosis of canine piroplasmids have been developed by Allsopp (1994), Ano *et al.*, (2001), Criado-Fornelio *et al.*, (2003a,b), Birkenheuer *et al.*, (2003) and also by other authors. Jefferies *et al.*, (2007) have reported a PCR-RFLP assay for canine piroplasmids.

The canine hepatozoons have been also diagnosed by standard PCR by Inokuma *et al.*, (2002); Criado-Fornelio *et al.*, (2003b) and Criado-Fornelio *et al.*, (2006). In addition, Criado-Fornelio *et al.*, (2007a) published a PCR-RFLP method for diagnosis of *Hepatozoon* spp.

There had been few attempts to develop standard PCR-based multiplex assays for multi-pathogen diagnosis in canids. Only Kledmanee *et al.*, (2009) described an assay for detection of *Ehrlichia*, *Babesia* and *Hepatozoon* in dogs.

2) Real-Time PCR (or Quantitative PCR, Abbreviated as qPCR)

Recently a SYBR green real-time High Resolution Melting (HRM) PCR assay for diagnosis of the *Babesia canis* complex species in dogs has been described by Adaszek and Winiarczyk (2010). The only procedure of this kind destined to diagnosis of *Hepatozoon* sp in canids was described by Criado-Fornelio *et al.*, (2007b). The same method was subsequently applied to cats with success (Criado-Fornelio *et al.*, 2009). Real-time PCR with FRET probes have been described both for piroplasmids (Wang *et al.*, 2010) and *Hepatozoon* (Li *et al.*, 2008). Peleg *et al.*, (2010) published a qPCR assay for multiplex detection of *Babesia canis vogeli* and *Ehrlichia canis* in dogs (by employing Taqman probes). Criado-Fornelio (2012a and b) has developed two separate FRET assays for simultaneous detection of hematozoans (piroplasmids and *Hepatozoon*). Both diagnostic tests have also been employed in DNA samples of Spanish canids (dog and red fox) with success (please see the "Epizootiology" section in this chapter).

3) Loop-Mediated Amplification (LAMP)

Müller *et al.*, (2010) diagnosed the three species included in the *Babesia canis* complex by this procedure. This method is quite interesting since the visual detection of amplified products circumvents the electrophoresis step of the standard PCR technique and moreover, it is relatively easy to perform, inexpensive and applicable to diagnosis in the field. In addition, it may be converted to the real-time PCR assay format (Criado-Fornelio, 2007).

4) Reverse Line Blotting (RLB)

Matjilla *et al.*, (2007, 2008a,b) employed RLB to diagnose the presence of piroplasmid infections in domestic dog and African wild dog (*Lycaon pictus*) in South Africa. Georges *et al.*, (2008) applied the same technique in domestic dogs in Trinidad. Jongehan *et al.*, (2011) studied the efficacy of an acaricide drug mix to prevent the tick attachment and the subsequent babesial infection (*B. canis canis*) in dogs from Holland by means of the RLB method. The blotting procedure is very interesting since it provides reliable identification of pathogens by employing oligonucleotidic probes. However, the RLB assay is relatively cumbersome and moreover, it requires well-trained personnel in a specialized laboratory (Criado-Fornelio, 2007).

5) In Situ Hybridization Assays

Yamasaki *et al.*, (2011) reported for the first time the development and validation of an *in situ* hybridization test for *B. gibsoni*. The parasite was detected by microscopic observations of thin smear samples. A digoxigenin-labelled DNA probe (based on the sequence of the Hsp70 gene of *B. gibsoni*) was used in this method. The technique worked well with either blood smears or paraffin-embedded tissues.

The FISH technique (a RNA-based method for detection of pathogens) has been used for diverse parasitic protozoa (Lucy *et al.*, 2008), but not for piroplasmids or *Hepatozoon*. However, some companies (IGeneX®, Palo Alto, CA, USA) are providing a private diagnosis service based on an unpublished FISH assay.

The technique allegedly detects *Babesia microti* and other *Babesia* species in humans. The possibility of applying such tests for veterinary diagnosis has not been envisaged up to now (Anonymous, 2011).

6) Miscelaneous Methods

Flow cytometry has been employed for detection of canine piroplasmids. The technical principle of this methodology is the use of hydroethidine, a compound that is absorbed by viable cells and then is converted (by cell metabolism) to ethidium bromide.

The latter compound binds to parasite DNA and provides a chemical tag to discriminate *Babesia*-infected erythrocytes. Diverse authors have succesfully assayed such technology to detect canine *Babesia* species (Uemura *et al.*, 1990; Fukata *et al.*, 1996; Bicalho *et al.*, 2004). An alternative method using a different nucleic acid stain (SYTO16) is also available for detection of *B. gibsoni* (Yamasaki *et al.*, 2008).

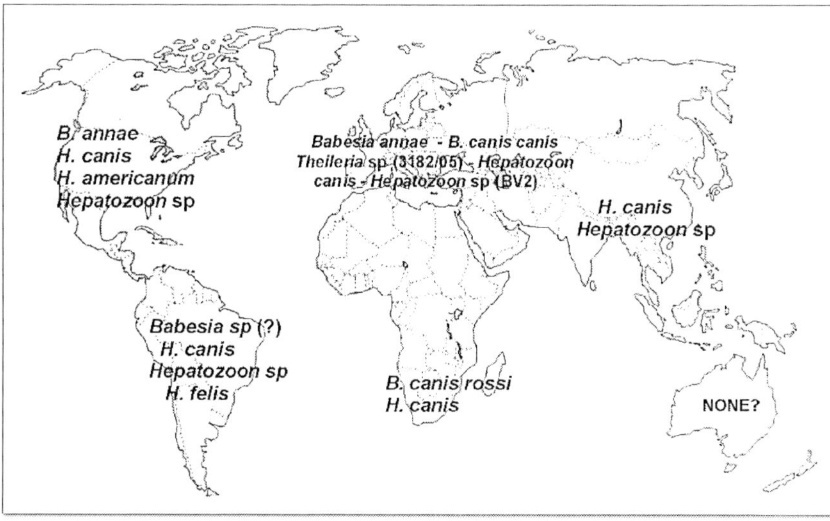

Figure 3. Approximate geographic distribution of the hematozoans of wild canids transmitted by ticks (piroplasmids and *Hepatoozon*). The tick-borne protozoa of the domestic dog have not been included in the map.

Attempts to apply an alternative microfluidic technology for diagnosis of bovine piroplasms have been published (Nascimento *et al.*, 2008). Nonetheless, it has not been employed for detection of canine hematozoans.

6. Epizootiology of Piroplasmids and Hepatozoons in Canids with Emphasis in Wildlife

Before the advent of DNA-based identification methods, the canine piroplasmids were limited to four species: *B canis canis*, *B. canis vogeli*, *B. canis rossi* and *B. gibsoni*. However, diverse studies have shown that there is a broader spectrum, which includes *Babesia* (*Theileria*) *annae*, *T. equi*, *T. annulata*, *Theileria* sp (diverse isolates related to at least two ruminant species), *B. caballi*, *B. conradae*, *Rangelia vitalii* and *Babesia* sp. The same is true for canine hepatozoons, where *H. canis* was thought to be the only species in dog and wild canids.

However, new hepatozoons have been described in dogs or foxes, such as *H. americanum*, *H. felis*, *Hepatozoon* sp (genotypes similar to isolate BV2 of a Spanish bank vole and others).

Table 3. Summary of published data on the presence of tick-borne protozoa in wild canids

Country / Diagnostic method	Host	Piroplasmid species / Prevalence (animals studied)	*Hepatoozon* spp / Prevalence (No. animals studied)	Publication
South Africa / Microscopy	African wild dog	*Babesia canis* / 7% (28)	*Hepatoozoon canis?* / 93% (28)	Van Heerden *et al.*, 1995
USA / Microscopy	Coyote	NA	*H. americanum* / 56% (16)	Kocan *et al.*, (1999)
USA / Microscopy	Coyote	NA	*H. americanum* / 40% (20)	Kocan *et al.*, (2000)
USA / PCR + S	Red fox	*Babesia annae* / NA (1)	NA	Goethert and Telford (2003)
USA / induced infection	Coyote	*B. gibsoni* / NA *(experimental infection)*	NA	Evers *et al.*, (2003)
Spain / PCR + S	Red fox	*B. annae* / 50% (10)	*H. canis* / 100 % (10)	Criado-Fornelio *et al.*, 2003b
Israel / ELISA	Red fox	NA	*H. canis* / 24 % (10)	Fishman *et al.*, 2004
Spain / PCR + S	Red fox South American foxes	NA NA	*H. canis* / 90% (20) *H. canis* and *H. americanum*-like/ 77% (13)	Criado-Fornelio et *al.*, (2006)
Spain / PCR + S	Red fox	*B. annae* / 50% (20)	*H. canis*/ 35% (20)	Criado-Fornelio *et al.*, (2007a)
Slovakia / PCR + S	Red fox	ND	*H. canis* / 44% (9)	Majlathova *et al.*, (2007)
Italy/ PCR + S	Red fox	*B. canis canis* / 22% (9) *B. annae* / 11% (9)	NA	Torina *et al.*, (2007)
South Africa / RLB	African wild dog	*Babesia canis* / 5% (301)	*Hepatoozoon canis* / 0.6% (301)	Matjilla *et al.*, (2008b)
Spain / PCR + S	Red fox	*B. annae* / 14% (5)	*H. canis* / 28% (5)	Giménez *et al.*, (2009)
Venezuela / PCR + S	Crab-eating fox	ND / (2)	ND / (2)	Criado-Fornelio *et al.*, (2009)
Croatia / PCR + S	Red fox	*B. annae* / 5% (191) *Theileria* sp 3182 05 / 0.5 % (191)	*H. canis* / 23 % (191) *Hepatozoon* sp BV-1 / 1 % (191)	Dezdek *et al.*, (2010)
Poland / PCR + S	Red fox	*B. annae* / 0.7 % (138)	*H. canis* / 11 % (138)	Karbowiak *et al.*, (2010)
Italy / PCR + S	Red fox	NA	*H. canis* / 13 % (119)	Gabrielli *et al.*, (2010)

Country / Diagnostic method	Host	Piroplasmid species / Prevalence (animals studied)	*Hepatoozon* spp / Prevalence (No. animals studied)	Publication
Brazil / PCR + S	2 bush dogs, 1 maned wolf and 2 South American foxes (*)	NA	*Hepatozoon sp* / 5% (100)	André *et al.*, 2010
Brazil / Serology	12 crab-eating foxes, 7 maned wolves and 2 hoary foxes	Not detected (21)	Not tested	Curi *et al.*, (2010)
USA / PCR + S	Coyote and grey fox	NA	*H. americanum* (coyote) /100% (3) *H. canis* (grey fox) / NA (1)	Allen *et al.*, (2011)
Brazil / Serology	Diverse South American wild canids	*Babesia* sp / 10% (94)	NA	André *et al.*, (2011)
India / PCR + S	Indian wild dog	NA	*H. canis* / 100 % (2)	Pawar *et al.*, (2012)
Argentina / PCR + S	South American fox	NA	*H. felis* / NA (1)	Giannitti *et al.*, (2012)
Spain / qPCR with FRET probes	Red fox	*B. annae* / 43% (30)	*H. canis* / 73 % (30)	Criado-Fornelio - present chapter

Emphasis is given to publications where diagnosis was based on molecular methods. Abbreviations: NA= Not Applicable; ND = not detected; PCR + S = Polymerase Chain Reaction and DNA sequencing; RLB = Reverse Line Blotting. (*) = only the 5 animals found infected are mentioned in the table; the remaining wild canids (n = 95) were negative.

Some of them have been described only at molecular level and their specific status is unclear at this moment.

Diverse recent publications have revised the distribution of the main tick-borne protozoa in the dog (Ahantarib, *et al.*, 2008; Irwin, 2009; Little *et al.*, 2009; Solano-Gallego and Baneth, 2011; Baneth, 2011). Murray *et al.*, (1999) revised the situation in wild canids but that work was published before genetic identification procedures were widely applied. Therefore, the information previously available is relevant but partial in what concerns the species present in wild canids and their impact on conservation. This section is aimed to fill in the gap in research advances on hematic protozoa of wild canids. As we will see, it is fairly evident that there had been few advances in Parasitology as compared to microbial diseases.

The distribution of the main species of piroplasmid and hepatozoons of wild canids is shown in Figure 3. In Europe (Table 3), it has become evident that at least one piroplasmid species is shared by dogs and red foxes in some countries. That is the case for *Babesia annae* in Spain, where the infection has been found both in dogs (Camacho *et al.*, 2001) and in foxes (Criado-Fornelio *et al.*, 2003b and 2007a; Giménez *et al.*, 2009). It is reported in this section a molecular survey for piroplasmids and *Hepatozoon* performed on DNA samples of 30 red foxes from Spain that had been obtained in a former study (Criado-Fornelio *et al.*, 2000). These wild canids had not been analysed in previous works (Criado-Fornelio *et al.*, 2003b; Criado-Fornelio *et al.*, 2006 and 2007a). The survey was completed with 100 blood samples obtained in stray dogs from a local animal shelter. The real-time PCR methods of Criado-Fornelio (2012, a and b) were employed for molecular diagnosis. In the group of 30 foxes, the two qPCR methods showed (by melting curve analysis) that 22 animals were infected by *Hepatozoon* sp. The employ of FRET probes HEF and HE705 (Criado-Fornelio, 2012b) demonstrated that the infecting species was *H. canis* (this specific identification for canine hepatoozons was not possible with FRET probes TAFA and TAFS - Criado-Fornelio, 2012a). In addition it was found that 13 foxes were infected by *B. annae*. In the group of 100 dogs, five animals were infected by *H. canis* (when employing probes HEF and HE705 as described by Criado -Fornelio, 2012b) and one animal was infected with *B. canis*. Protozoan isolates from foxes were not sequenced. However, sequencing was performed in the samples of positive dogs. The *B. canis* isolate (which could not be adscribed to any subspecies by melting curve analysis) was identified as *B. canis vogeli*. In the 5 dogs infected by *Hepatozoon* (99.4 to 100 % identity to the *H. canis* isolate Spain-2 – AY461378), some slight sequence polymorphisms were observed. Mutations were found at position 19 (T → G) in one dog isolate and positions 196 and 228 (in both cases G → A) in another dog isolate. These differences may have been due to sequencing errors or strain variation. Therefore, the Spanish isolates were identified as *H. canis*, which is in agreement with the identification based on the melting curve analysis in assays with probes HEF and HE705. The results reported in this chapter confirmed that the recently published methods of Criado-Fornelio (2012a and b) may be helpful in the diagnosis of both feline and canine hematozoans. In addition these data suggest that in Spain the genotype Spain-1 is typical of foxes whereas the genotype Spain-2 is the most common in dogs. On the other hand, diverse data obtained in European countries (Croatia), show that *B. annae* is present both in red foxes (5%) and dogs (3%) as demonstrated by Dezdek *et al.*, (2010) and

Beck *et al.*, (2009). In Poland, *B. annae* is present as well in foxes with low frequency (0.7% - Karbowiak *et al.*, 2010), whereas it was not detected in a small group of foxes in Slovakia (Majlathova *et al.*, 2007). In a molecular survey conducted in a small group of foxes (9 animals) in Italy, Torina *et al.*, (2007) diagnosed infections caused by *B. canis canis* and *B. annae*. In addition, the finding of a Croatian red fox infected by *Theileria* sp (related to roe deer isolate 3182/05) was reported by Votja *et al.*, (2009).

Concerning canine hepatozoonosis in Europe, *H. canis* was frequently found in European red foxes by molecular methods in Spain (Criado-Fornelio *et al.*, 2003b, 2006, 2007a and data in the present chapter), Slovakia (Majlathova *et al.*, 2007), Poland (Karbowiak *et al.*, 2010), Croatia (Dezdek *et al.*, 2010) and Italy (Gabrielli *et al.*, 2011). In all those countries prevalence ranged from 73-100% (in Spain) to 11% (in Poland). It is interesting to mention that an isolate of *Hepatozoon* sp (related to *Hepatozoon* sp BV2 of a Spanish rodent) has been found with low prevalence both in red fox and dog in Croatia (Vojta *et al.*, 2009; Dezdek *et al.*, 2010). These findings suggest that the prey-predator transmission cycles may be common for certain *Hepatozoon* species (Allen *et al.*, 2011). Since rabbits are one of the main food sources for red foxes in the Central Iberian Peninsule (Rosalino *et al.*, 2007; Delibes-Mateos *et al.*, 2008), it is not strange that surveys conducted in Central Spain have failed to detect *Hepatozoon* species of rodents (Criado-Fornelio *et al.*, 2003b and 2006; Giménez *et al.*, 2009 and Criado-Fornelio data in this chapter). Instead, red foxes from northern Spain usually capture rodents (Hernández, 2005). Therefore, it is tempting to speculate that these northern populations might harbor rodent hepatozoons (as it does occur in dogs in Croatia).

In North America, *Babesia annae* has been found in the red fox (Goethert and Telford, 2003; Clancey *et al.*, 2010) but not in dog. There are no reports of other piroplasms in wild canids, except the successful experimental *B. gibsoni* infection in coyote carried out by Evers *et al.*, (2003). In the USA, *H. canis* has been found in grey foxes and *H. americanum* in coyotes (Allen *et al.*, 2011). Previous reports already indicated that *H. americanum* is relatively common in coyotes (Kocan *et al.*, 1999 and 2000). *H. canis* seems to be present in dogs in North America, but it is far less frequent than *H. americanum* (Allen *et al.*, 2008; Li *et al.*, 2008).

In South America there are limited data on piroplasmids present in wild canids. In general they have not been encountered in the few molecular surveys carried out up to now (Criado-Fornelio *et al.*, 2006; André *et al.*, 2010; Curi *et al.*, 2010). In one serological survey in Brazil (André *et al.*,

2011), it was found that *Babesia* sp antibodies were present in 10% of the wild canids analysed, but DNA amplification did not reveal any piroplasmid infection in these animals. The situation is quite different for *Hepatozoon. H. canis*, *H. felis* and *Hepatozoon* sp have been reported in South American foxes by Criado-Fornelio *et al.*, (2006 and 2009), André *et al.*, (2010), Giannitti *et al.*, (2012). Some of the Brazilian hepatozoons seem to be loosely related to *H. americanum* (Criado-Fornelio *et al.*, 2006). In contrast, dogs in South America are mainly infected by *H. canis* (Criado-Fornelio *et al.*, 2007b and 2009; Rubini *et al.*, 2008; Spolidorio *et al.*, 2009). These findings suggest that transmission cycles linked to ingestion of the definitive host (plus predation or vertical transmission) have been far more successful than vectorial transmission linked to blood feeding, at least in the American continent. Alternative explanations are possible, indeed. For instance, some wild canids species may act as dead-end hosts for piroplasmids, the same as may occur for some viral pathogens (Bolzoni *et al.*, 2012). Further research is needed to address such hypotheses.

In Africa, molecular epizootiology studies concerning piroplasmosis and hepatozoonosis of wild canids have been limited to South Africa. *B. canis rossi* has been found in jackals (Nuttal, 1912), although it was not diagnosed by molecular methods. In addition, infection caused by *B. canis rossi* and/or *H. canis* (Van Heerden *et al.*, 1995; Matjilla *et al.*, 2008b; Penzhorn, 2011) has been reported in African wild dog. Much more attention has been paid to tick-borne protozoa of African wild felines due to the existence of endangered feline species (Munson *et al.*, 2008). At present, there is no information pointing to the presence of *B. annae* in carnivores in Africa.

There had been little new significant data on the molecular epizootiology of tick-borne hematozoans of wild canids in Asia. The only recent work in this field has been reported by Pawar *et al.*, (2012), who found that Indian wild dogs were infected by *H. canis*. The same species was found in foxes and dogs in Japan (Maede *et al.*, 1982; Inokuma *et al.*, 2002) and also in dogs in India (Pawar *et al.*, 2012; Abd-Rani *et al.*, 2011), Thailand (Jittapalapong *et al.*, 2006; Criado-Fornelio *et al.*, 2007a) and the Philippines (Baticados *et al.*, 2011).

Table 4 shows the arthropod hosts for piroplasmids and hepatozoons of wild canids. Some of the vectors listed are hypothetical, as only indirect supporting evidence has been presented so far. It is interesting to point out that in some instances the spectrum of putative vectors is continuosly changing. In fact, some tick-borne hematozoans show wider host range that previously suspected (e.g. *H. canis* and *B. annae*).

Table 4. Summary of known and putative arthropod vectors for the most relevant piroplasmids and hepatozoons of wild canids

Hematozoan species	Arthropod vector	Type of evidence	Publication(s)
Babesia annae	*Ixodes hexagonus*, *Rhipicephalus* spp , *Ixodes* spp	Indirect	Camacho *et al.*, 2003; Cassini *et al.*, (2009); Iori *et al.*, 2010; Lledó *et al.*, 2010,
B. canis rossi	*Haemaphysalis elliptica*	Direct	Horak *et al.*, (1987); Horak, (1995); Penzhorn, (2011)
Hepatozoon canis	*Rhipicephalus sanguineus, R. microplus, Amblyomma ovale*	Direct	Baneth *et al.*, (2007); Forlano *et al.*, (2005) ; Rubini *et al.*, (2009) ; De-Miranda *et al.*, (2011)
H. felis	*Rhipicephalus sanguineus*	Indirect	Criado-Fornelio, (2012b)
H. americanum	*Amblyomma maculatum*	Direct	Ewing *et al.*, (2002 and 2003)

This applies both to intermediate and definitive host. Thus, *Amblyomma* ticks (*A. ovale*) may be able to transmit *H. canis,* according to Forlano *et al.*, (2005 and 2008). In addition, it is very likely that diverse *Rhipicephalus* species may transmit *H. canis* infection, as reported in South America (Criado-Fornelio *et al.*, 2007a; De-Miranda *et al.*, 2011). Rosalino *et al.*, (2007) did not detect specimens of *R. sanguineus* in red fox scat in Portugal, but *R. pusillus* and *R. turanicus* instead.

Some *Ixodes* species were also detected (*e.g. I. hexagonus*, *I. ventalloi*, *I. ricinus*). Hence, ingestion of *R. sanguineus* ticks by red foxes seems to be uncommon in that country. Interestingly, *H. canis* has been diagnosed with high frequency (48%) in wild red foxes in Portugal (Conceiçao-Silva *et al.*, 1988). Altough these data do not prove in anyway transmission of *H. canis* by multiple tick species, the elucidation of the role played by species other than *R. sanguineus* in hematozoan transmission in Europe certainly deserves further investigation.

7. Evolution and Phylogeny of Piroplasmids and *Hepatozoon* Species of Canids

Data currently available suggest that *Babesia annae* is a piroplasmid that first evolved in carnivore mammals (albeit ungulates should not be totally disregarded) from Eurasia. These hosts may have contributed to spread the disease to some regions of North America. The lack of appropriate vectors may have precluded the transmission of *B. annae* to carnivores in South America, thus explaining the absence of the piroplasmid in that part of the world (as well as in Africa). However, there are alternative explanations for the relatively restricted geographic range of *B. annae*, such as the insufficient number of wild carnivores sampled both in South America and Africa.

Recently Penzhorn (2011) speculated with the possibility that *B. canis vogeli* (or a related species) originated in wolf and then it passed to domestic dog. This was based in the idea that it is the least virulent *B. canis* subspecies, therefore it should have been for a long time associated with domestic dogs. However, it seems much more likely that *B. canis vogeli* originated in cat, since it is much less pathogenic in that host, as shown by Simking *et al.*, (2010). If the pathogenicity criterium is crucial when considering the evolution of parasites, the reverse trend (*i.e.* from canids to felids) may have occurred for the subspecies *B. canis canis/presentii.*

It is evident that these parasites induce noticeable disease signs in felines (Criado-Fornelio *et al.*, 2003a; Baneth *et al.*, 2004). The fourth subspecies in the *B. canis* complex (*B. canis rossi*) is likely to have evolved in African wild canids (Penzhorn, 2011).

In infections caused by that piroplasmid in jackals and African wild dogs, disease signs are infrequent. However, it is interesting to underline the fact that this subspecies has been diagnosed in asymptomatic dogs and horses from France (Fritz, 2010). If these infections are occurring in other European countries, the evolutionary scheme might be somewhat different.

In this chapter we summarize the evolution of canine *Hepatozoon* spp, in correspondence with the mammalian carnivore lineage (Criado-Fornelio, 2012b). The complete 18S rRNA gene sequences of *Babesia annae* (outgroup), *Hepatozoon* sp (bank vole 1), *H. felis* and several *H. canis* sequences (isolates Spain 1, Spain 2 and Curupira 1, 2 and 3) were aligned with ClustalW, and then analyzed with the MEGA 5 software to obtain a evolutionary history tree, which was based on the molecular clock hypothesis (Figure 4).

The evolutionary history was inferred by using the Maximum Likelihood method based on the Kimura 2-parameter model. The landmark of separation of *Hepatozoon* from rodents and carnivores is dated at the Cretaceous–Tertiary (K-T) boundary (62 million years ago = MYA), as shown by the fossil record (Springer *et al.*, 2003).

Such time landmark was introduced in the MEGA5 software for dating other nodes in the tree. In summary, the results of the analysis indicate that:

- *H. felis* and the canid hepatozoon lineage diverged approximately 48 MYA (node 1).
- *H. americanum* and *H. canis* separated about 39 MYA (node 2).
- *H. canis* from dogs and European foxes separated 12 MYA (node 3)
- *H. canis* from European and South American foxes separated 6 MYA (node 4).

These data are more or less in agreement with the evolution of vertebrate hosts based on the fossil record or on studies employing on DNA sequence analysis. Thus, the last common ancestor of living carnivores may have existed as recently as 42 MYA (Polly *et al.*, 2006). However, other reports indicate that a common ancestor was already present about 55 MYA (Springer *et al.*, 2003). Therefore, the average of both hypotheses is 55 + 42 / 2 = 48.5, which is in quite good agreement with the timing obtained for node number 1.

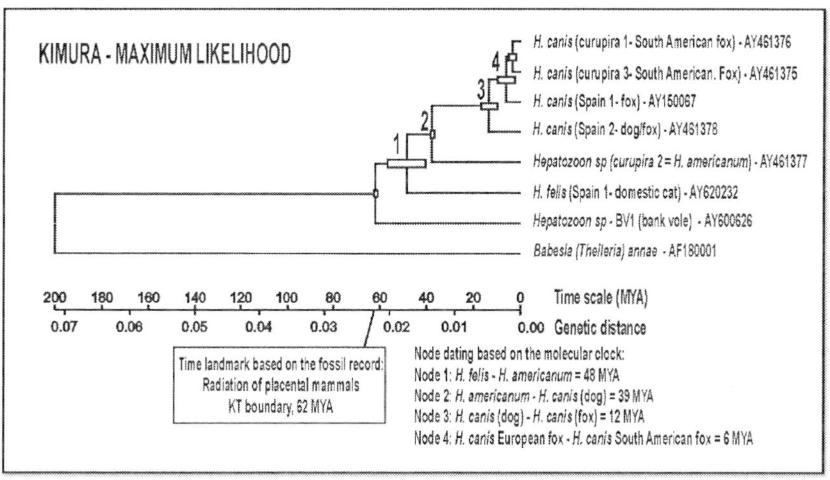

Figure 4. Phylogenetic relationships of *Hepatozoon* spp and *Babesia annae* based on DNA sequence variation of the 18S rRNA gene (MEGA 5 program). The Genbank® accession numbers are indicated after each name. The scale below indicates both time and genetic distance. The evolutionary history was inferred by using the Maximum Likelihood method (based on the Kimura two-parameter model). See the text of the chapter for more details on the analysis.

The available epizootiological data suggest that *H. felis* is clearly a parasite typical of felines (Criado-Fornelio, 2012b) that has been seldom found in canids (Giannitti *et al.*, 2012).

The dating for node number 2 is 39 MYA, which could be correlated more or less with the first canines dated in the American fossil record (34 MYA), according to Tedford *et al.*, (2009). The dating for node number 3 is 12 MYA, corresponding to the dichotomy of *Hepatozoon* from dogs (genotype Spain 2) and foxes (genotype Spain 1). This is in very good agreement with the timing of separation of foxes and dogs according to Tedford *et al.*, (2009). The hypothesis that the genotype Spain 1 is more frequent in foxes (Criado-Fornelio *et al.*, 2006) has been confirmed by studies carried out in this and other European countries (Criado-Fornelio, 2003b; Majlathova et al., 2007; Giménez *et al.*, 2009; Gabrielli *et al.,* 2011). This argues against the hypothesis proposed by Little *et al.*, (2009), who suggested that canine *Hepatozoon* species may be a complex of related genogroups rather than well defined species. This may be the case when dog isolates only are considered. However, it is evident that phylogenetic analysis combined with the finding of partial host specificity (as may be the case with isolates found in dog and red foxes in Spain) reveals a certain degree of divergency. If these data justify the erection of separated species is a controversial taxonomic issue that shall not be discussed here.

The node number 4 may be correlated with the separation of European and South American foxes, which dates back to 6 MYA according to the molecular clock hypothesis, and 5 MYA based on the fossil record (Tedford *et al.*, 2009). This idea is based on the assumption that the genotype Curupira 3 is more frequent in wild canids in South American, since it has not been found in European foxes up to now (Criado-Fornelio *et al.*, 2006 and 2007a; Majlathova *et al.*, 2007; Giménez *et al.*, 2009; Karbowiak *et al.*, 2010; Dezdek *et al.*, 2010; Gabrielli *et al.,* 2011). It is important to point out that in many publications, sequences obtained in *Hepatozoon* spp isolates are based on partial 18S rRNA gene sequences (exactly in the central part of the gene). Therefore, there is some degree of uncertainity in the molecular identifications performed. This is due to the fact that the 5' end of the small ribosomal subunit gene is considered the most suitable region for identification purposes ("DNA barcoding") in many organisms (Bhadury *et al.*, 2006).

Additional evidence for the relative accuracy of molecular dating is based on the fact that according to this method, the separation of *Theileria annae* and *Hepatozoon* occurred in the Jurassic period, 200 MYA (see Figure 4). This is in agreement with the idea that ticks appeared during the Mesozoic Era, as pointed out by Balashov (1989). Thus, it seems that in the primitive Apicomplexan arthropod-borne group, the divergence between two possible modes of transmission (one linked to ingestion, another to blood feeding) occurred in the early Mesozoic.

8. Conservation Medicine and Control of Tick-Borne Hematozoans of Wild Canids

Tick-Borne Hematozoans and Conservation Medicine

Wild canids may serve as excellent sentinel species of emerging canine tick-borne diseases. A sentinel species is defined as one that provides insight into the ecosystem status (Aguirre, 2009). The ecosystem condition is assessed by monitoring diverse variables (including diseases, parasites, toxics, climatic change, habitat destruction, etc). In this sense, emerging parasitic tick-borne diseases have influence on wildlife conservation and moreover, they may also affect both domestic animals and human public health. In the case of animals, it is fairly evident that pathogen-derived mortality in wild canids is much more likely to be caused by microbial parasites (*e.g.* rabies or canine distemper virus) than by parasitic protozoa (Murray *et al.*, 1999). The impact of hematozoans on their hosts depends on the infecting species, the severity of infection, the possible concurrent pathogens and some host-related factors (such as age, immune status and reproductive status - Giannitti *et al.*, 2012). Likewise, the health and nutritional status of the host indirectly influences the level of infection. Another factor that influences the parasitic burden is the density of population. Crowding facilitates the spread of parasites with direct life cycles, but the effect is lesser in parasites with indirect life cycles. Most wild canid species are not endangered so that the risk of elimination of populations due to lethal protozan infections is remote. Nonetheless, is quite difficult to assess the extent of harm caused by parasites in nature. Data obtained on parasitic protozoa of wild canids in the past have been limited and often sporadic. In addition, animals killed by pathogen infection are rarely found and recovered. To further complicate this issue, wild canid populations usually show low densities, have secretive habitats and they usually are afraid of humans (Michalsiki *et al.*, 2011). Hence, the actual mortality ratio caused by most parasites in nature is largely unknown (Malan *et al.*, 2007). Moreover, studies on the pathogenic effects caused by parasites to wild animals (along with analysis of their influence on host vitality, reproduction, and population density) are limited and sometimes contradictory (Balashov, 2007). Published studies on standard hematic and biochemical parameters in wild canids are limited (Aguirre, 2009). As a consequence, the assessment of parasite-induced changes in wild canid groups is a tricky task. Diminution of canid populations may occur in outbreaks of parasitic disease, but usually there is a recovery soon afterwards (Begon and Bowers, 1995). The problem may be much more complicated if wild canids live near to a dense population of reservoir hosts (*e.g.* domestic dog – Bronson *et al.*, 2008).

Wild canid species that are not threatened by extinction are not always protected by conservation agents. Thus, in some European countries, red fox (*Vulpes vulpes*) populations have been limited by hunting or trapping. This is due to the fact that foxes may feed on small game species (or protected species), negatively affecting the local economy or the conservation of other wildlife species (Ruiz-Olmo *et al.*, 1990).

The development of new research tools such as molecular detection methods has prompted new insights in studies on wildlife parasites (Thompson *et al.*, 2010). An example of this ongoing revolution (although not related to wild canids) is the finding that simians are the origin of some human malaria parasites (*Plasmodium falciparum*). This was inferred by molecular analysis performed on gorilla faeces (Liu *et al.*, 2010). Such approach has been poorly employed for pathogens of wild canids, but it looks less conflicting than the capture of wild animals in order to obtain blood samples. In fact, faeces are abundant and probably the most collectable animal product in the wild and may be acquired without disturbance to the study animals (Wasser *et al.*, 2004). Both the parasite and host species may be identifed in fecal material by molecular procedures. The system is ideal for temperate countries, but it may work well even for samples obtained in the tropics, where exposition to humidity, warm temperatures, rain and sunlight might rapidly degrade DNA (Michalski *et al.*, 2011). With this system, it has been possible to study parasitic fauna (helmints or ticks) in wildlife (Rosalino *et al.*, 2006) and in the red fox (Rosalino *et al.*, 2007). The method has not been employed for hematic parasites up to now, but this possibilty should not be disregarded

Wild Canids and Parasite Control Issues

There is a commercial vaccine for *B. canis canis* in dogs, but its effects on wild canids have not been tested. In contrast, there are no vaccines available for canine hepatozoonosis (Bowman *et al.*, 2001; Baneth, 2011). One of the problems associated with designing effective vaccines is that hematozoans display mechanisms that efficiently evade host immunity. In addition, hematic protozoa show noticeable species diversity and/or strain variability. Furthermore, little is known about how dogs generate a protective immunity against these pathogens (Shaw *et al.*, 2001).

Preventive measures may be adopted against the arthropod vector, although their application may not be feasible in the habitat of wild canids. Tick infestations are common and elimination can be expensive and time consuming. Effective tick control is difficult to achieve, because of the abundance of potential alternative hosts in the tick life cycle. Eradication is not feasible except on isolated locations such as islands. The use of acaricides in the environment (where tick infection usually occurs) is not exempt of risks (toxicity, residues in animal products and pollution). Treatment on domestic animals by pour-on or spot-on procedures is therefore preferred (Jongehan and Uilenberg, 1994). Many effective tick control agents exist, several of which also possess activity against other parasites (Blagburn and Dryden, 2009). Tick vaccines are complicated in the case of *Rhipicephalus* species, since these are frequently three-host ticks and immunization of reservoir hosts may not be feasible.

Tick-borne parasites of dogs may be treated with diverse antiprotozoal drugs. For those interested in the treatment of domestic canids, the reading of the reviews by Irwin, (2009), Baneth (2011) and Solano-Gallego and Baneth (2011) is recommended. Chemotherapy of intracellular protozoa is complicated since drug action is limited in that particular environment. Sometimes treatment leads to clinical cure but the animal may still harbour parasites (the "Trojan horse" reservoir, as defined by Baneth *et al.*, 2012). The application of combined drug therapy may help to overcome this problem. Sometimes different drugs are required for treating different parasite species (Shaw *et al.*, 2001). An additional problem in the treatment of canine hepatozoonosis is the existence of tisular stages that are barely affected by chemotherapy (Lapin, 2010 and Baneth, 2011). Moreover, there is little experience in wild canid chemotherapy. It is likely that drugs employed for treatment of babesiosis and/or hepatozoonosis in dogs (such as imidocarb dipropionate, atovacuone, toltrazuril and others) may be parasiticidal in related canines as well. However, the possible secondary effects of these drugs on wild animals are largely unknown.

Usually the control of parasites by individual treatment/vaccination of wildlife species is unfeasible, due to the dispersal of individuals in a wide geographic area and the problems associated to handling wild canid species. In general, in many natural parks, very seldom will veterinarians intervene, since such action is incongruent with the principles of nature conservation (Malan *et al.*, 1997). However some authors have suggested that chemotherapy may be an option to avoid the spreading of helminthic parasitic diseases (such as echinococcosis) in wild red foxes (Ito *et al.*, 2003).

CONCLUSION

Tick-borne hematozoans such as piroplasmids and *Hepatozoon* spp are nowadays seen as emergent pathogens in wildlife, including canids. Many new hematozoans have been discovered since the arrival of molecular diagnosis methods. The presence of such a diverse parasitic fauna in wild canids emphasizes the need of further investigations on vectors and reservoir hosts. In general, tick-borne protozoa do not threat endangered canine species, albeit the number of parasitological surveys carried out is insufficient to claim that there is no risk at all. Problems may be exacerbated due to extrinsic stress factors (such as climatic changes associated to global warming). The presence of concomitant pathogens (mostly viral diseases) may further complicate the survival of endangered canid species. The possibility of controlling parasites in the wild does not seem feasible. Moreover, since transmission of some hematic protozoa (*e.g.* hepatozoons) might be linked to predation, preventive measures are very difficult to undertake.

All these facts emphasize the urgent need of investigating the epizootiology of tick-borne protozoa in extremely endangered canid species. The lack of data is more evident in Asia, where much more additional reseach is needed. The use of scat surveys is highly recommended for such purposes. However, before these analyses start, it is neccesary that preliminary studies determine whether molecular analysis of fecal material is a method reliable for detection of hematic parasites or not.

ACKNOWLEDGMENTS

I am deeply indebted with Dr Catalina Rey-Valeirón for her critical reading of the manuscript and helpful suggestions. I also wish to thank the veterinarians of the SMP animal shelter for their invaluable help in the collection of animal samples.

REFERENCES

Abd-Rani, P.A., Irwin, P.J., Coleman, G.T., Gatne, M., Traub, R.J., (2011). A survey of canine tick-borne diseases in India. *Parasites and Vectors*, 4:141.

Aboge, G.O., Jia, H., Kuriki, K., Zhou, J., Nishikawa, Y., Igarashi, I., Fujisaki, K., Suzuki, H., Xuan, X., (2007). Molecular characterization of a novel 32-kDa merozoite antigen of *Babesia gibsoni* with a better diagnostic performance by enzyme-linked immunosorbent assay. *Parasitology*, 134, 1185-1194.

Adachi, K., Yoshimoto, A., Hasegawa, T., Shimizu, T., Goto, Y., Makimura, S., (1992). Anti-erythrocyte membrane antibodies detected in sera of dogs naturally infected with *Babesia gibsoni*. *Journal of Veterinary Medicine Science*, 54, 1081-1084.

Adaszek, C., Winiarczyk, S., (2010). Application of the SYBR green real-time HRM PCR technique in the differentiation of the *Babesia canis canis* protozoa isolated in the areas of eastern Poland. *Parasitology Research*, 106, 1253-1256.

Aguirre, A., (2009). Wild canids as sentinels of ecological health: a conservation medicine perspective. *Parasites and Vectors*, 2S1:S7.

Ahantarib, A., Trindachartvanit, N., Milne, J.R., (2008). Tick-born pathogens and diseases of animals and humans in Thailand. *Southeast Asian Journal of Tropical Medicine and Public Health*, 39, 1015-1032.

Allen, K.E., Li, B., Kaltenboek, E.M., Johnson, E.M., Reichard, M.V., Panciera, R.J., Little, S.E., (2008). Diversity of *Hepatozoon* species in naturally infected dogs from the southern United States. *Veterinary Parasitology*, 154, 220-225.

Allen, K.E., Yabsley, M.J., Johnson, E.M., Reichard, M.V., Panciera, R.J., Ewing, S.A., Little, S.E., (2011). Novel *Hepatozoon* in vertebrates from the southern United States. *Journal of Parasitology*, 97, 648-653.

Allsopp, M.T., Cavalier-Smith, T., De Waal, D.T., Allsopp, B.A., (1994). Phylogeny and evolution of the piroplasms. *Parasitology*, 108, 147-152.

Anderson, R.M., May, R.M., (1978). Regulation and stability of host-parasite population interactions. I. Regulatory processes. *Journal of Animal Ecology*, 47, 219-247.

André, M.R., Adania, C.H., Teixeira, R.H., Vargas, G.H., Falcade, M., Sousa, L., Salles, A.R., Allegretti, S.M., Felippe, P.A., Machado, R.Z., (2010). Molecular detection of *Hepatozoon* spp. in Brazilian and exotic wild carnivores. *Veterinary Parasitology*, 173, 134-138.

André, M.R., Adania, C.H., Rodrigo Teixeira, H.F., Allegretti, S.M., Machado, R.Z., (2011). Molecular and serological detection of *Babesia* spp. in neotropical and exotic carnivores in Brazilian Zoos. *Journal of Zoo and Wildlife Medicine*, 42, 139-143.

Ano, H., Makimura, S., Harasawa, R., (2001). Detection of *Babesia* species from infected dog blood by polymerase chain reaction. *Journal of Veterinary Medical Science*, 63, 111-113.

Anonymous, (2011). IGeneX Inc, Palo Alto, CA. Quality Assurance and Proficiency Statement. Available at: http://igenex.com/files/QA_PACKAGE_2010.pdf.

Balashov, I.S., (1989). The coevolution of ixodid ticks and terrestrial vertebrates. *Parazitologiia*, 23, 457-468.

Balashov, I.S. (2007). Harmfulness of Parasitic Insects and Acarines to Mammals and Birds. *Entomological Review*, 87, 1300–1316.

Baneth, G., Harmelin, A., Presentey, B.Z., (1995). *Hepatozoon canis* infection in two dogs. *Journal of the American Veterinary Medical Association*, 206, 1891-1894.

Baneth, G., Breitschwerdt, E.B., Hegarty, B.C., Pappalardo, B., Ryan, J. (1998a). A survey of tick-borne bacteria and protozoa in naturally exposed dogs from Israel. *Veterinary Parasitology*, 74, 133-142.

Baneth, G., Shkap, V., Samish, M., Pipano, E., Savitsky, I., (1998b). Antibody response to *Hepatozoon canis* in experimentally infected dogs. *Veterinary Parasitology*, 74, 299-305.

Baneth, G., Shkap, V., (2003). Monozoic cysts of *Hepatozoon canis*. *The Journal of Parasitology*, 89, 379-381.

Baneth, G., Kewnny, M.J., Tasker, S., Anug, Y., Shkap, V., Levy, A., Shaw, S.E., (2004). Infection with a proposed new subspecis of *Babesia canis*, *Babesia canis* subsp. *presentii*, in domestic cats. *Journal of Clinical Microbiology*, 42, 99-105.

Baneth, G., Samish, M., Shkap, V., (2007). Life cycle of *Hepatozoon canis* (Apicomplexa: Adeleorina: Hepatozoidae) in the tick *Rhipicephalus sanguineus* and domestic dog (*Canis familiaris*). *Journal of Parasitology*, 93, 283-299.

Baneth, G., (2011). Perspectives on canine and feline hepatozoonosis. *Veterinary Parasitology*, 181, 3-11.

Baneth, G., Bourdeau, P., Bourdoiseau, G., (2012). Vector-borne diseases – constant challenge for practicing veterinarians: recommendations from the CVBD World Forum. *Parasites and Vectors*, 5:55.

Baticados, A.G., Baticados, W.N., Carlos, E.T., Carlos, S.M., Villarba, L.A., Subiaga, S.G., Magcalas, J.M., (2011). Parasitological detection and molecular evidence of *Hepatozoon canis* in canids from Manila, Philippines. *Veterinary Medicine: Research and Reports*, 2010, 7-10.

Beck, R., Vojta, L., Mrljak, V., Marinculić, A., Beck, A., Zivicnjak, T., Cacciò, S.M., (2009). Diversity of *Babesia* and *Theileria* species in symptomatic and asymptomatic dogs in Croatia. *International Journal for Parasitology*, 39, 843-848.

Begon, M., Bowers, R.G., (1995). *Ecology of Infectious diseases in natural populations*. Grenfell B.T., and Dobson, A.P., (Eds.); Cambridge University Press, Cambridge (UK).

Bhadury, P., Austen, M., Bilton, D., Lambshead, P.J., Rogers, A., Smerdon, G., (2006). Development and evaluation of a DNA-barcoding approach for the rapid identification of nematodes. *Marine Ecology Progress Series*, 320, 1-9.

Bicalho, K.A., Ribeiro, M.F., Martins-Filho, O.A., (2004). Molecular fluorescent approach to assessing intraerythrocytic hemoprotozoan *Babesia canis* infection in dogs. *Veterinary Parasitology*, 125, 221-235.

Birkenheuer, A.J., Levy, M.G., Breitschwerdt, E.B., (2003). Development and evaluation of a seminested PCR for detection and differentiation of *Babesia gibsoni* (Asian genotype) and *B. canis* DNA in canine blood samples. *Journal of Clinical Microbiology*, 41, 4172-4177.

Birkenheuer, A.J., Horney, B., Bailey, M., Scott, M., Sherbert, B., Catto, V., Marr, H.S., Camacho, A.T., Ballman, A.E., (2010). *Babesia microti*-like infections are prevalent in North American foxes. *Veterinary Parasitology*, 172, 179-182.

Blagburn, B.L, Dryden, M.W., (2009). Biology, treatment, and control of flea and tick infestations. *Veterinary Clinics North American Small Animals Practitioner*, 39, 1173-1200.

Bolzoni, L., Rosa, L., Cagnacci, F., Rizzoli A., (2012). Effect of dear density on tick infestation of rodents and the hazard of tick-borne encephalitis. Part II. Population and infection models. *International Journal for Parasitology*, 42, 365-372.

Bowman, D.D., Hendrix, D.M., Lindsay, D.S., Barr, S.C., (2001). Hemogregarines. *In: Feline Clinical Parasitology*. Ed, Iowa State University Press, Ames, Iowa, pp 41-42.

Bronson, E., Emmons, L.H., Murray, S., Dubovi, E.J., Deem ,S.L., (2008). Serosurvey of pathogens in domestic dogs on the border of Noël Kempff Mercado National Park, Bolivia. *Journal of Zoo and Wildlife Medicine*, 39, 28-36.

Camacho, A.T., Pallas, E., Gestal, J.J., Guitián, F.J., Olmeda, A.S., Goethert, H.K., Telford, S.R., (2001). Infection of dogs in north-west Spain with a *Babesia microti*-like agent. *The Veterinary Record*, 149, 552-555.

Camacho, A.T., Pallas, E., Gestal, J.J., Guitián, F.J., Olmeda, A.S., Telford, S.R., Spielman, A., (2003). *Ixodes hexagonus* is the main candidate as vector of *Theileria annae* in northwest Spain. *Veterinary Parasitology*, 112, 157-163.

Cardinal, M.V., Reitniger, R., Gürtler, R., (2006). Use of an immunochromatographic dipstick test for rapid detection of *Trypanosoma cruzi* in sera from animal reservoir hosts. *Journal of Clinical Microbiology*, 44, 3005-3007.

Cassini, R., Zanutto, S., Frangipane-di-Regalbono, A., Gabrielli, S., Calderin, P., Moretti, A., Tampieri, M.P., Pietrobelli, M., (2009). Canine piroplasmosis in Italy: epidemiological aspects in vertebrate and invertebrate hosts. *Veterinary Parasitology*, 165, 30-35.

Clancey, N., Horney, B., Burton, S., Birkenheuer, A., McBurney, S., Tefft, K., (2010). *Babesia* (*Theileria*) *annae* in a red fox (*Vulpes vulpes*) from Prince Edward Island, Canada. *Journal of Wildlife Diseases*, 46, 615-621.

Colly, L.P., Nesbit, J.W., (1992). Fatal acute babesiosis in a juvenile wild dog. *Journal of the South African Veterinary Association*, 63, 36-38.

Conceiçao-Silva, F.M., Abranches, P., Silva-Pereira, M.C., Janz, J,G., (1988). Hepatozoonosis in foxes from Portugal. *Journal of Wildlife Diseases*, 24, 344-347.

Criado-Fornelio, A., Gutierrez-García, L., Rodríguez-Caabeiro, F., Reus-García, E., Roldán-Soriano, M.A., Díaz-Sánchez, M.A., 2000. A parasitological survey of wild red foxes (*Vulpes vulpes*) from the province of Guadalajara, Spain. *Veterinary Parasitology*, 92, 245-251.

Criado-Fornelio, A., Martinez-Marcos, A., Buling-Saraña, A., Barba-Carretero, J.C., (2003a). Presence of *Mycoplasma haemofelis*, *Mycoplasma haemominutum* and piroplasmids in cats from southern Europe: a molecular study. *Veterinary Microbiology*, 93, 307-317.

Criado-Fornelio, A., Martinez-Marcos, A., Buling-Saraña, A., Barba-Carretero, J.C., (2003b). Molecular studies on *Babesia*, *Theileria* and *Hepatozoon* in southern Europe. Part I. Epizootiological aspects. *Veterinary Parasitology*, 113, 189-201.

Criado-Fornelio, A., Martinez-Marcos, A., Buling-Saraña, A., Barba-Carretero, J.C., (2003c). Molecular studies on *Babesia*, *Theileria* and *Hepatozoon* in southern Europe. Part II. Phylogenetic analysis and evolutionary history. *Veterinary Parasitology*, 114, 173-194.

Criado-Fornelio, A., Ruas, J.L., Casado, N., Farias, N.A., Soares, M.P., Müller, G., Brumt, J.G., Berne, M.E., Buling-Saraña, A., Barba-Carretero, J.C., (2006). New molecular data on mammalian *Hepatozoon* species (Apicomplexa: Adeleorina) from Brazil and Spain. *Journal of Parasitology*, 92, 93-99.

Criado-Fornelio, A., (2007). A review of nucleic acid-based diagnostic tests for *Babesia* and *Theileria*, with emphasis on bovine piroplasms. *Parassitologia*, 49S1, 39-44.

Criado-Fornelio, A., Rey-Valeiron, C., Buling, A., Barba-Carretero, J.C., Jefferies, R., Irwin, P., (2007a). New advances in molecular epizootiology of canine hematic protozoa from Venezuela, Thailand and Spain. *Veterinary Parasitology*, 144, 261-269.

Criado-Fornelio, A., Buling, A., Cunha-Filho, N.A., Ruas, J.L., Farias, N.A., Rey-Valeiron, C., Pingret, J.L., Etievant, M., Barba-Carretero, J.C., (2007b). Development and evaluation of a quantitative PCR assay for detection of *Hepatozoon* sp. *Veterinary Parasitology*, 150, 352-356.

Criado-Fornelio, A., Buling, A., Casado, N., Gimenez, C., Ruas, J.L., Wendt, L., Farias, N.A., Pinheiro, M., Rey-Valeiron, C., Barba-Carretero, J.C., (2009). Molecular characterization of arthropod-borne hematozoans in wild mammals from Brazil, Venezuela and Spain. *Acta Parasitologica*, 54, 187-193.

Criado-Fornelio, A., (2012a, in press). Emerging tick-borne protozoal diseases - part 1. New advances in diagnosis, epizootiology and taxonomy of feline piroplasmids. In: *Felines: behavior, classification and diseases*. Editors Moreno-García, R.E. and Ruiz-Jiménez, R.G.. Nova Science Publishers, Hauppage, New York, USA.

Criado-Fornelio, A., (2012b, in press). Emerging tick-borne protozoal diseases - part 2. Feline hepatozoonosis and the "resurrection" of *Hepatozoon felis*. In: *Felines: behavior, classification and diseases*. Editors Moreno-García, R.E. and Ruiz-Jiménez, R.G.. Nova Science Publishers, Hauppage, New York, USA.

Curi, N.H.D.A., Araújo, A.S., Campos, F.S., Lobato, Z.I.P., Gennari, S.M., Marvulo, M.F.V., Silva, J.C.R., Talamoni, S.A., (2010). Wild canids, domestic dogs and their pathogens in Southeast Brazil: disease threats for canid conservation. *Biodiversity and Conservation*, 19, 3513-3524.

Da-Silva, A.S., França, R.T., Costa, M.M., Paim, C.B., Paim, F.C., Dornelles, G.L., Soares, J.F., Labruna, M.B., Mazzanti, C.M., Monteiro, S.G., Lopes, S.T., (2011). Experimental infection with *Rangelia vitalii* in dogs: acute phase, parasitemia, biological cycle, clinical-pathological aspects and treatment. *Experimental Parasitology*, 128, 347-352.

Day, M.J., (2011). The immunopathology of canine vector-borne diseases. *Parasites and Vectors*, 13, 48.

Delibes-Mateos, M., Fernández De Simon, J., Villafuerte, R., Ferreras, P. (2008). Feeding responses of the red fox (*Vulpes vulpes*) to different wild rabbit (*Oryctolagus cuniculus*) densities: A regional approach. *European Journal of Wildlife Research*, 54, 71-78.

De-Miranda, R.L., de Castro, J.R., Olegário, M.M., Beletti, M.E., Mundim, A.V., O'Dwyer, L.H., Eyal, O., Talmi-Frank, D., Cury, M.C., Baneth, G., (2011). Oocysts of *Hepatozoon canis* in *Rhipicephalus* (*Boophilus*) *microplus* collected from a naturally infected dog. *Veterinary Parasitology*, 177, 392-396.

Dezdek, D., Vojta, L., Curković, S., Lipej, Z., Mihaljević, Z., Cvetnić, Z., Beck, R., (2010). Molecular detection of *Theileria annae* and *Hepatozoon canis* in foxes (*Vulpes vulpes*) in Croatia. *Veterinary Parasitology*, 172, 333-336.

Dunn, R.R., Harris, N.C., Colwell, R.K., Koh, L.P., Sodhi, N.S., (2009). The sixth mass coextinction: are most endangered species parasites and mutualists? *Proceedings of Biological Sciences*, 276, 3037-3045.

Du Plessis, J.L., Fourie, N., Nel, P.W., Evezard, D.N., (1990). Concurrent babesiosis and ehrlichiosis in the dog: blood smear examination supplemented by the indirect fluorescent antibody test, using *Cowdria ruminantium* as antigen. *Onderstepoort Journal of Veterinary Research*, 57, 151-155.

East, M.L., Wibbelt, G., Lieckfeldt, D., Ludwig, A., Goller, K., Wilhelm, K., Schares, G., Thierer, D., Hofer, H., (2008). A *Hepatozoon* species genetically distinct from *H. canis* infecting spotted hyenas in the Serengeti ecosystem, Tanzania. *Journal of Wildlife Diseases*, 44, 45-52.

Eizirik, E., Murphy, W.J., Koepfli, K.P., Johnson, W.E., Dragoo, J.W., Wayne, R.K., O'Brien, S.J., (2010). Pattern and timing of diversification of the mammalian order Carnivora inferred from multiple nuclear gene sequences. *Molecular Phylogenetics and Evolution*, 56, 49-63.

Evers, H.V., Kocan, A.A., Reichard, M.V., Meinkhot, J,H., (2003). Experimental *Babesia gibsoni* infection in coyotes (*Canis latrans*). *Journal of Wildlife Diseases*, 39, 904-908.

Ewing, S.A, Matthew, J.S., Panciera, R.J., (2002). Transmission of *Hepatozooon americanum* (Apicomplexa: Adeleorina) by ixodids (Acari: Ixodidae). *The Journal of Medical Enomology*, 39, 631-634.

Ewing, S.A, Panciera, R.J., (2003). American canine hepatozoonosis. *Clinical Microbiology Reviews*, 16, 688-697.

Fishman, Z., Gonen, L., Harrus, S., Strauss-Ayali, D., King, R., Baneth, G., (2004). A serosurvey of *Hepatozoon canis* and *Ehrlichia canis* antibodies in wild red foxes (*Vulpes vulpes*) from Israel. *Veterinary Parasitology*, 119, 21-26.

Forlano, M., Scofield, A., Elisei, C., Fernandes, K.R., Ewing, S.A., Massard, C.L., (2005). Diagnosis of *Hepatozoon* spp. in *Amblyomma ovale* and its experimental transmission in domestic dogs in Brazil. *Veterinary Parasitology*, 134, 1-7.

Forlano, M., Mujica, F., Coronado, A., Meléndez, R.D., Linardi, P.M., Botelho, J.R., Bellosta, P., Barrios, N., (2008). Especies de *Amblyomma* (acari: ixodidae) parasitando perros (*Canis familiaris*) en áreas rurales de los estados Lara, Yaracuy, Carabobo y Falcón, Venezuela. *Revista Científica, FCV-LUZ* , 18, 662 – 666.

Fritz D., (2010). A PCR study of piroplasms in 166 dogs and 111 horses in France (March 2006 to March 2008). *Parasitology Research*, 106, 1339-1142.

Fukata, T., Ohnishi, T., Okuda, S., Sasai, K., Baba, E., Arakawa, A., (1996). Detection of canine erythrocytes infected with *Babesia gibsoni* by flow cytometry. *Journal of Parasitology*, 82, 641-642.

Fukumoto, S., Xuan, X., Inoue, N., Igarashi, I., Sugimoto, C., Fujisaki, K., Nagasawa, H., Mikami, T., Suzuki, H.., (2003). Molecular characterization of a gene encoding a 29-kDa cytoplasmic protein of *Babesia gibsoni* and evaluation of its diagnostic potentiality. *Molecular and Biochemical Parasitology*, 131, 129-136.

Fukumoto, S., Sekine, Y., Xuan, X., Igarashi, I., Sugimoto, C., Nagasawa, H., Fujisaki, K., Mikami, T., Suzuki, H., (2004). Serodiagnosis of canine *Babesia gibsoni* infection by enzyme-linked immunosorbent assay with recombinant P50 expressed in *Escherichia coli*. *Journal of Parasitology*, 90, 387-391.

Furuta, P.I., De Sousa-Oliveira, T.M., Alves-Teixeira, M.C., Rocha, A.G:, Machado, R.Z., Tinucci-Vosta, M., (2009). Comparison between a soluble antigen-based ELISA and IFAT in detecting antibodies against *Babesia canis* in dogs. *Revista Brasileira de Parasitologia Veterinaria*, 18, 41-45.

Gabrielli, S., Kumlien, S., Calderini, P., Brozzi, A., Iori, A., Cancrini, G., (2010). The first report of *Hepatozoon canis* identified in *Vulpes vulpes* and ticks from Italy. *Vector Borne and Zoonotic Diseases*, 10, 855-9.

García, A.T., (2006). Piroplasma infection in dogs in northern Spain. *Veterinary Parasitology*, 138, 97-102.

Gaunt, P.S., Gaunt, S.D., Craig, T.M., (1983). Extreme neutrophilic leukocytosis in a dog with hepatozoonosis. *Journal of the American Veterinary Medical Association*, 182, 409-410.

Georges, K., Ezeokoli, C.D., Newaj-Fyzul, A., Campbell, M., Mootoo, N., Mutani, A., Sparagano, O.A., (2008). The application of PCR and reverse line blot hybridization to detect arthropod-borne hemopathogens of dogs and cats in Trinidad. *Annals of the New York Academy of Sciences*, 1149, 196-199.

Georgieva, Y., Konthur, Z., (2011). Design and screening of M13 phage display cDNA libraries. *Molecules*, 16, 1667-1681.

Giannitti, F., Diab, S.S., Uzal, F.A., Fresneda, K., Rossi, D., Talmi-Frank, D., Baneth, G., (2012) , Infection with a *Hepatozoon* sp. closely related to *Hepatozoon felis* in a wild Pampas gray fox (*Lycalopex -Pseudalopex - gymnocercus*) co-infected with canine distemper virus. *Veterinary Parasitology*, 186, 497-502.

Giménez, C., Casado, N., Criado-Fornelio, A., de Miguel, F.A., Domínguez-Peñafiel, G., (2009). A molecular survey for Piroplasmida and *Hepatoozon* in wild animals in Burgos (Northern Spain). *Veterinary Parasitology*, 162, 147-150.

Goethert, H.K., Telford, R., (2003). What is *Babesia microti*? *Parasitology*, 127, 301-309.

Gonen, L., Strauss-Ayali, D., Shkap, V., Vincent-Johnson, N., Macintire, D.K., Baneth G., (2004). An enzyme-linked immunosorbent assay for antibodies to *Hepatozoon canis*. *Veterinary Parasitology*. 122, 131-139.

Goo, Y.K., Jia, H., Aboge, G.O., Terkawi, M.A., Kuriki, K., Nakamura, C., Kumagai, A., Zhou, J., Lee, E.G., Nishikawa, Y., Igarashi, I., Fujisaki, K., Xuan, X., (2008). *Babesia gibsoni*: Serodiagnosis of infection in dogs by an enzyme-linked immunosorbent assay with recombinant BgTRAP. *Experimental Parasitology*, 118, 555-560.

Goo, Y.K., Jia, H., Terkawi, M.A., Aboge, G.O., Yamagishi, J., Nishikawa, Y., Kim, S., Jang, H.K., Fujisaki, K., Xuan, X., (2009a). *Babesia gibsoni*: identification, expression, localization, and serological characterization of a *Babesia gibsoni* 22-kDa protein. *Experimental Parasitology*, 123, 273-276.

Goo, Y.K., Jia, H., Aboge, G.O., Terkawi, M.A., Lee, E.G., Yamagishi, J., Nishikawa, Y., Jang, H.K., Sunaga, F., Namikawa, K., Fujisaki, K., Xuan, X., (2009b). Characterization of the *Babesia gibsoni* 12-kDa protein as a potential antigen for the serodiagnosis. *Parasitology International*, 58, 55-60.

Gray, A.D., Akin, M.T., McLean, R., Valeri, C.R., (1991). Evaluation of the Quantitative Buffy Coat (QBC) method to detect malaria-infected red blood cells. *Milestones in Medicine*, 156, 241-5.

Hänscheid, T., (1999). Diagnosis of malaria: a review of alternatives to conventional microscopy. *Clinical Laboratory of Haematology*, 21, 235-245.

Harrus, S., Baneth, G., (2005). Drivers for the emergence and re-emergence of vector-borne protozoal and bacterial diseases. *International Journal for Parasitology*, 35,1309-1318.

Hernández, A., (2005). Topos en la dieta de zorros rojos en el noroeste de España. *Galemys*, 17, 87-90.

Hirose, Y., Matsuhira, S., Batanova, T.A., Goo, Y.K., Terkawi, M.A., Nishikawa, Y., Kitoh, K., Xuan, X., Takashima, Y., (2009). Detection of *Babesia gibsoni* by reaction of phage display single chain antibodies with P50 proteins. *Journal of Veterinary Medicine Science*, 71, 1133-1136.

Horak, I.G., Jacot-Guillarmod, A., Moolman, L.C., de Vos, V. (1987). Parasites of domestic and wild animals in South Africa. XXI. Ixodid ticks on domestic dogs and on wild carnivores. *Onderstepoort Journal of Veterinary Research*, 54, 573-580.

Horak, I.G., (1995). Ixodid ticks collected at the Faculty of Veterinary Science, Onderstepoort, from dogs diagnosed with *Babesia canis* infection. *Journal of the South African Veterinary Association*, 66, 170-171.

Hosse, R.J., Rothe, A., Power, B.E., (2006). A new generation of protein display scaffolds for molecular recognition. *Protein Science*, 15, 14-27.

Inokuma, H., Okuda, M., Ohno, K., Shimoda, K., Onishi, T., (2002). Analysis of the 18S rRNA gene sequence of a *Hepatozoon* detected in two Japanese dogs. *Veterinary Parasitology*, 106, 265-271.

Iori, A., Gabrielli, S., Calderini, P., Moretti, A., Pietrobelli, M., Tampieri, M.P., Galuppi, R., Cancrini, G., (2010). Tick reservoirs for piroplasms in central and northern Italy. *Veterinary Parasitology*, 170, 291-296.

Irwin, P.J., (2009). Canine babesiosis: from molecular taxonomy to control. *Parasites and Vectors*, 2:S4.

Ito, A., Romig, T., Takahashi, K., (2003). Perspective on control options for *Echinococcus multilocularis* with particular reference to Japan. *Parasitology*, 127, S159-172.

IUCN 2011. IUCN (International Union for Conservation of Nature and Natural Resources) Red List of Threatened Species. Version 2011.2. <www.iucnredlist.org>.

Jefferies, R., Ryan, U.M., Irwin, P.J., (2007). PCR-RFLP for the detection and differentiation of the canine piroplasm species and its use with filter paper-based technologies. *Veterinary Parasitology*, 144, 20-27.

Jia, H., Zhou, J., Ikadai, H., Matsuu, A., Suzuki, H., Igarashi, I., Fujisaki, K., Xuan, X., (2006). Identification of a novel gene encoding a secreted antigen 1 of *Babesia gibsoni* and evaluation of its use in serodiagnosis. *American Journal of Tropical Medicine and Hygiene*, 75, 843-850.

Jia, H., Liao, M., Lee, E., Nishikawa, Y., Inokuma, H., Ikadai, H., Matsuu, A., Igarashi, I., Xuan, X., (2007). Development of an immunochromatographic test with recombinant BgSA1 for the diagnosis of *Babesia gibsoni* infection in dogs. *Parasitology Research*, 100, 1381-1384.

Jia, H., Terkawi, M.A., Aboge, G.O., Goo, Y.K., Zhou, J., Lee, E.G., Nishikawa, Y., Igarashi, I., Fujisaki, K., Xuan, X.., (2008). *Babesia gibsoni*: identification of an immunodominant, interspersed repeat antigen. *Experimental Parasitology*, 118, 146-149.

Jia, H., Terkawi, M.A., Aboge, G.O., Goo, Y.K., Ma, L., Zhou, J., Nishikawa, Y., Igarashi, I., Fujisaki, K., Xuan, X., (2009). Identification of secreted antigen 3 from *Babesia gibsoni. Clinical Vaccine Immunology*, 16, 944-948.

Jittapalapong, S., Rungphisutthipongse, O., Maruyama, S., Schaefer, J.J., Stich, R.W., (2006). Detection of *Hepatozoon canis* in stray dogs and cats in Bangkok, Thailand. *Annals of the New York Academy of Sciences*, 1081, 479-488.

Jongehans, F., Uilenberg, G. (1994). Ticks and control methods. *Revue Scientifique et Technique – Office International des Epizooties*, 13, 1201-1226.

Jongejan, F., Fourie, J.J., Chester, S.T., Manavella, C., Mallouk, Y., Pollmeier, M.G., Baggott, D., (2011). The prevention of transmission of *Babesia canis canis* by *Dermacentor reticulatus* ticks to dogs using a novel combination of fipronil, amitraz and (S)-methoprene. *Veterinary Parasitology*, 179, 343-350.

Karbowiak, G., Majlathova, V., Hapunik, K., Peyko, B., Wita, I., (2010). Apicomplexan parasites of red foxes (*Vulpes vulpes*) in Northeastern Poland. *Acta Parasitologica*, 55, 204-214.

Kjemtrup, A.M., Wainwright, K., Miller, M., Penzhorn, B.L., Carreño, R.A.(2006). *Babesia conradae*, sp. Nov., a small canine *Babesia* identified in California. *Veterinary Parasitology*, 138, 103-111.

Kledmanee, K., Suwanpakdee, S., Krajangwong, S., Chatsiriwech, J., Suksai, P., Suwannachat, P., Sariya, L., Buddhirongawatr, R., Charoonrut, P., Chaichoun, K., (2009). Development of multiplex polymerase chain reaction for detection of *Ehrlichia canis*, *Babesia* spp and *Hepatozoon canis* in canine blood. *Southeast Asian Journal of Tropical Medicine and Public Health*, 40, 35-39.

Kocan, A.A., Breshears, M., Cummings, C., Panciera, R.J., Ewing, S.A., Barker, R.W., (1999). Naturally occurring hepatozoonosis in coyotes from Oklahoma. *The Journal of Wildlife Diseases*, 35, 86-89.

Kocan, A.A., Cummings, C.A., Panciera, R.J., Mathew, J.S., Ewing, S.A., Barker, R.W., (2000). Naturally occurring and experimentally transmitted *Hepatozoon americanum* in coyotes from Oklahoma. *The Journal of Wildlife Diseases*, 36, 149-153.

Kumar, M., Shekhar, P., Haque, S., Mahto, D., (2008). Feline babesiosis. *Veterinary World*, 1, 120-121.

Lapin, M.R., (2010). Update on the diagnosis and management of *Hepatozoon* spp infections in the USA. *Top Companion Animal Medicine*, 25, 142-144.

Li, Y., Wang, C., Allen, K.E., Little, S.E., Ahluwalia, S.K., Gao, D., Macintire, D.K., Blagburn, B.L., Kaltenboeck, B., (2008). Diagnosis of canine *Hepatozoon* spp. infection by quantitative PCR. *Veterinary Parasitology*, 157, 50-58.

Little, S.E., Allen, K.E., Johnson, E.M., Panciera, R.J., Reichard, M.V., Ewing, S.A., (2009). New developments in canine hepatozoonosis in North America: a review. *Parasites and Vectors*, 2S1: S5.

Liu, W., Li, Y., Learn, G.H., Rudicell, R.S., Robertson, J.D., Keele, B.F., Ndjango, J.B., Sanz, C.M., Morgan, D.B., Locatelli, S., Gonder, M.K., Kranzusch, P.J., Walsh, P.D., Delaporte, E., Mpoudi-Ngole, E., Georgiev, A.V., Muller, M.N., Shaw, G.M., Peeters, M., Sharp, P.M., Rayner, J.C., Hahn, B.H., (2010). Origin of the human malaria parasite *Plasmodium falciparum* in gorillas. *Nature*, 467, 420-425.

Lledó, L., Giménez-Pardo, C., Domínguez-Peñafiel, G., Sousa, R., Gegúndez, M.I., Casado, N., Criado-Fornelio, A., (2010). Molecular detection of hemoprotozoa and *Rickettsia* species in arthropods collected from wild animals in the Burgos Province, Spain. *Vector Borne and Zoonotic Diseases*, 10, 735–738.

Loretti A.P., Barros, S.S., (2005). Hemorrhagic disease in dogs infected with an unclassified intraendothelial piroplasm in southern Brazil. *Veterinary Parasitology*, 134, 193-213.

Lucy, F.E., Graczyk, T.K., Tamang, L., Miraflor, A., Minchin, D., (2008). Biomonitoring of surface and coastal water for *Cryptosporidium*, *Giardia*, and human-virulent microsporidia using molluscan shellfish. *Parasitology Research*, 103, 1369-1375.

Maede, Y., Ohsugi, T., Ohothasi, N., (1982). *Hepatozoon* infection in a wild fox (*Vulpes vulpes* schrecenki Kishida) in Japan. *Nihon Juigaku Zasshi*, 44, 137-142.

Majláthová, V., Hurníková, Z., Majláth, I., Petko, B., (2007). *Hepatozoon canis* infection in Slovakia: imported or autochthonous? *Vector Borne and Zoonotic Diseases*, 7, 199-202.

Malan, F.S., Horak, I.G., De Vos V., Van Wyk J.A., (1997). Wildlife parasites: Lessons for parasite control in livestock. *Veterinary Parasitology*, 71, 137-153.

Mathew, J.S., Saliki, J., Ewing, S.A., Lehenbauer, T.W., Panciera, R.J., Malayer, J.R., Cummings, C.A., Kocan, A.A., (2001). An indirect enzyme-linked immunosorbent assay for diagnosis of American canine hepatozoonosis. *Journal of Veterinary Diagnostic Investigations*, 13, 17-21.

Matjila, P.T., Penzhorn, B.L., Leisewitz, A.L., Bhoora, R., Barker, R., (2007). Molecular characterisation of *Babesia gibsoni* infection from a pit-bull terrier pup recently imported into South Africa. *Journal of the South African Veterinary Association*, 78, 2-5.

Matjila, P.T., Leisewitz, A.L., Oosthuizen, M.C., Jongejan, F., Penzhorn, B.L., (2008a). Detection of a *Theileria* species in dogs in South Africa. *Veterinary Parasitology*, 157, 34-40.

Matjila, P.T., Leisewitz, A.L., Jongejan, F., Bertschinger, H.J., Penzhorn, B.L., (2008b). Molecular detection of *Babesia rossi* and *Hepatozoon* sp. in African wild dogs (*Lycaon pictus*) in South Africa. *Veterinary Parasitology*, 157, 123-127.

McCully, R.M., Basson, P.A., Bigalke, R.D., De Vos, V., Young, E., (1975). Observations on naturally acquired hepatozoonosis of wild carnivores and dogs in the Republic of South Africa. *Onderstepoort Journal of Veterinary Research*, 42, 117-133.

Mercer, S.H., Craig, T.M., (1988). Comparison of various staining procedures in the identification of *Hepatozoon canis* gamonts. *Veterinary Clinics Pathology*,17, 63-65.

Mercer, S.H., Jones, L.P., Rappole, J.H., Twedt, D., Laack, L.L., Craig, T.M., (1988). *Hepatoozon* sp in wild carnivores in Texas. *The Journal of Wildlife Diseases*, 24, 574-576.

Michalski, F., Valdez, F.P., Norris, D., Zieminski, C., Kashivakura, C.K., Trinca, C.S., Smith, H.B., Vynne, C., Wasser, S.K., Metzger, J.P., Eizirik, E., (2011). Successful carnivore identification with faecal DNA across a fragmented Amazonian landscape. *Molecular Ecology Resources*, 11, 862-871.

Miyama, T., Inokuma, H., Itamoto, K., Okuda, M., Verdida, R.A., Xuan, X., (2006). Clinical usefulness of antibodies against *Babesia gibsoni* detected by ELISA with recombinant P50. *Journal of Veterinary Medicine Science*, 68, 1371-1373.

Müller, H., Aysul, N., Liu, Z., Salih, D.A., Karagenc, T., Beyer, D., Kullmann, B., Ahmed, J.S., Seitzer, U., (2010). Development of a loop-mediated isothermal amplification (LAMP) assay for rapid diagnosis of *Babesia canis* infections. *Transbound Emergent Diseases*, 57, 63-65.

Munson, L., Terio, K.A., Kock, R., Mlengeya, T., Roelke, M.E., Dubovi, E., Summers, B., Sinclair, A.R., Packer, C., (2008). Climate extremes promote fatal co-infections during canine distemper epidemics in African lions. *PLoS One*, 3(6):e2545.

Murata, T., Inoue, M., Tateyama, S., Taura, Y., Nakama, S. (1993). Vertical transmission of *Hepatozoon canis* in dogs. *The Journal of Veterinary Medical Science*, 55, 867-868.

Murray, D.L., Kapke, C.A., Evermann, J.F., Fuller, T.K. (1999). Infectious disease and the conservation of free-ranging large Carnivores. *Animal Conservation*, 2, 241–254.

Namikawa, K., Sunaga, F., Kanno, Y., (1988). Morphology of *Babesia gibsoni* in canine erythrocytes. *Nihon Juigaku Zasshi*, 50, 936-938.

Nascimento, E.M., Nogueira, N., Silva, T., Braschler, T., Demierre, N., Renaud, P., Oliva, A.G., (2008). Dielectrophoretic sorting on a microfabricated flow cytometer: label free separation of *Babesia bovis* infected erythrocytes. *Bioelectrochemistry*, 73, 123-128.

Narantsatsral, S., Goo, Y.K., Battsetseg, B., Myagmarsuren, P., Terkawi, M.A., Soma, T., Luo, Y., Li, Y., Cao, S., Yu, L., Kamyingkird, K., Aboge, G.O., Nishikawa, Y., Xuan, X., (2011). Expression of truncated *Babesia gibsoni* thrombospondin-related adhesive proteins in *Escherichia coli* and evaluation of their diagnostic potential by enzyme-linked immunosorbent assay. *Experimental Parasitology*, 129, 196-202.

Noya, O., Alarcón de Noya, B., (1998). The multiple antigen blot assay (MABA): a simple immunoenzymatic technique for simultaneous screening of multiple antigens. *Immunology Letters*, 63, 53-56.

Nuttall, G.H.F., (1912). Note on *Rossiella rossi* (Nuttall, 1910) occurring in the jackal in British East Africa. *Parasitology*, 5, 61-64.

O'Dwyer, L.H., (2011). Brazilian canine hepatozoonosis. *Revista Brasileira de Parasitologia Veterinaria*, 20, 181-193.

Otranto, D., Testini, G., Dantas-Torres, F., Latrofa, M.S., Diniz, P.P., de Caprariis, D., Lia, R.P., Mencke, N., Stanneck, D., Capelli, G., Breitschwerdt, E.B., (2010). Diagnosis of canine vector-borne diseases in young dogs: a longitudinal study. *Journal of Clinical Microbiology*, 48, 3316-3324.

Panciera, R.J., Mathew, J.S., Cummings, C.A., Duffy, J.C., Ewing, S.A., Kocan, A.A., (2001). Comparison of tissue stages of *Hepatozoon americanum* in the dog using immunohistochemical and routine hystologic methods. *Veterinary Pathology*, 28, 422-426.

Pau, S., Leoni, A., Arru, E., Valenza, F., Guarda, F., (1982). Anatomia patologica dell'infezione da *Hepatozoon canis*. *Schweizer Archives für Tierheilkunde*, 124, 517-527.

Pawar, R.M., Poornachandar, A., Srinivas, P., Rao, K.R., Lakshmikantan, U., Shivaji, S., (2012). Molecular characterization of *Hepatozoon* spp. infection in endangered Indian wild felids and canids. *Veterinary Parasitology*, 186, 475-479.

Peleg, O., Baneth, G., Eyal, O., Inbar, J., Harrus, S., (2010). Multiplex real-time qPCR for the detection of *Ehrlichia canis* and *Babesia canis vogeli*. *Veterinary Parasitology*, 173, 292-299.

Penzhorn, B. (2006). Babesiosis of wild carnivores and ungulates. *Veterinary Parasitology*, 138, 11-21.

Penzhorn, B.L., (2011). Why is Southern African canine babesiosis so virulent? An evolutionary perspective. *Parasites and Vectors*, 4:51.

Polly, P.D., Wesley-Hunt, G.D., Heinrich R.E, Davis, G., Houde, P., (2006). Earliest known carnivoran auditory bulla and support for a recent origin of crowngroup Carnivora (Eutheria, Mammalia). *Palaeontology*, 49, 1019–1027.

Rosalino, L.M., Torres, J., Santos-Reis, M., (2006). A survey of helminth infection in Eurasian badger (*Meles meles*) in relation to their foraging behaviour in a Mediterranean environment in South west Portugal. *European Journal of Widlife Research*, 52, 202-206.

Rosalino, L.M., Rodrigues, M., Santos-Silva, M., Santos-Reis, M., (2007). Unusual findings on host-tick interactions through carnivore scat analysis. *Experimental and Applied Acarology*, 43, 293-302.

Rubini, A.S., dos Santos-Paduan, K., Von Ah Lopes, V., O'Dwyer L.H., (2008). Molecular and parasitological survey of *Hepatozoon canis* (Apicomplexa: Hepatozoidae) in dogs from rural area of Sao Paulo state, Brazil. *Parasitology Research*, 102, 895-899.

Rubini, A.S., Paduan, K.S., Martins, T.F., Labruna, M.B., O'Dwyer, L.H., (2009). Acquisition and transmission of *Hepatozoon canis* (Apicomplexa: Hepatozoidae) by the tick *Amblyomma ovale* (Acari: Ixodidae). *Veterinary Parasitology*, 164, 324-327.

Ruiz-Olmo, J., Grau, J. M. T., Puig, R. (1990). Comparación de la evolución de las poblaciones de zorro (*Vulpes vulpes* L., 1758) en el NE ibérico en base a datos históricos (siglos 18-19) y actuales (siglo 20). *Miscellania Zoologica*, 14, 225-231.

Schetters, T.P., Moubri, K., Cooke, B.M., (2009). Comparison of *Babesia rossi* and *Babesia canis* isolates with emphasis on effects of vaccination with soluble parasite antigens: a review. *Journal of the American Veterinary Association*, 80, 75-78.

Schmidt, G.D., Roberts, L.S. (1981). *Foundations of Parasitology*, 2nd Ed., C.V. Mosby Co. Eds., St Louis, Missouri, USA.

Shaw, S.E., Day, M.J., Birtles, R.J., Breitschwerdt, E.B., (2001). Tick-borne infectious diseases of dogs. *Trends in Parasitology*, 17, 74-80.

Shulkina, E.E., Beier, T.V., Sergiev, V.P., Lastrebova, R.L., (2004). Detection of hemogregarin of the genus *Hepatoozon* in patient in Russia (in Russian). *Medical Parasitology (Moscow)*, 4, 3-6.

Smith, T. G. 1996. The genus *Hepatozoon* (Apicomplexa: Adeleina). *Journal of Parasitology*, 82, 565-585.

Soares, J.F., Girotto, A., Brandão, P.E., Da-Silva, A.S., França, R.T., Lopes, S.T., Labruna, M.B., (2011). Detection and molecular characterization of a canine piroplasm from Brazil.*Veterinary Parasitology*, 180, 203-208.

Solano-Gallego, L., Baneth, G. (2011). Babesiosis in dogs and cats - expanding parasitological and clinical spectra. *Veterinary Parasitology*, 181, 48-60.

Spolidorio, M.G., Labruna, M.B., Zago, A.M., Donatele, D.M., Caliari, K.M., Yoshinari, N.H., (2009). *Hepatozoon canis* infecting dogs in the State of

Espírito Santo, southeastern Brazil. *Veterinary Parasitology*, 163, 357-361.

Springer, M.S., Murphy, W.J., Eizirik, E., O'Brien, S.J., (2003). Placental mammal diversification and the Cretaceous-Tertiary boundary. *Proceedings of the National Academy of Sciences of the U.S.A.*, 100, 1056-1061.

Suksawat, J., Xuejie, Y., Hancock, S.I., Hegarty, B.C., Nilkumhang, P., Breitschwerdt, E.B., (2001). Serologic and molecular evidence of coinfection with multiple vector-borne pathogens in dogs from Thailand. *Journal of Veterinary Internal Medicine*, 15, 453-462.

Tedford, R.H., Wang, X., Taylor, B.E., (2009). Phylogenetic Systematics of the North American Fossil Caninae (Carnivora: Canidae). *Bulletin of the American Museum of Natural History*, 325, 1-218.

Terkawi, M.A., Amornthep, A., Ooka, H., Aboge, G., Jia, H., Goo, Y.K., Nelson, B., Yamagishi, J., Nishikawa, Y., Igarashi, I., Kawazu, S.I., Fujisaki, K., Xuan, X., (2009). Molecular characterizations of three distinct *Babesia gibsoni* rhoptry-associated protein-1s (RAP-1s). *Parasitology*, 136, 1147-1160.

Thompson, R.C., Lymbery, A.J., Smith. A. (2010). Parasites, emerging disease and wildlife conservation. *International Journal for Parasitology*, 40, 1163-1170.

Torina, A., Vicari, D., Lo Valvo, M., Agnone, A., Reale, S., Alongi, A., Ferrantelli, V., (2007). Research on *Babesia* parasites in Sicilian wild red foxes. *Parassitologia*, 49S1, 108.

Trapp, S.M., Dagnone, A.S., Vidotto, O., Freire, R.L., Amude, A.M., de Morais, H.S., (2006). Seroepidemiology of canine babesiosis and ehrlichiosis in a hospital population. *Veterinary Parasitology*, 140, 223-230.

Uemura, T., Suzuki, S., Ohnishi, T., (1990). Flow cytometric enumeration of reticulocyte in the peripheral blood from canine infected with *Babesia gibsoni*. *Zentralblatt für Veterinarmedizin B*, 37, 468-472.

Van Heerden, J., Mills, M.G., Van Vuuren, M.J., Kelly, P.J., Dreyer, M.J., (1995). An investigation into the health status and diseases of wild dogs (*Lycaon pictus*) in the Kruger National Park. *Journal of the South African Veterinary Association*, 66, 18-27.

Verdida, R.A., Hara, O.A., Xuan, X., Fukumoto, S., Igarashi, I., Zhang, S., Dong, J., Inokuma, H., Kabeya, H., Sato, Y., Moritomo, T., Maruyama, S., Claveria, F., Nagasawa, H., (2004). Serodiagnosis of *Babesia gibsoni* infection in dogs by an improved enzyme-linked immunosorbent assay with recombinant truncated P50. *Journal of Veterinary Medicine Science*, 66, 1517-21.

Verdida, R.A., Xuan, X., Fukumoto, S., Huang, X., Zhou, J., Igarashi, I., Claveria, F.G., Nagasawa, H., (2005). Development of a practical immunochromatographic test with recombinant P50 for the diagnosis of *Babesia gibsoni* infection in dogs. *Parasitology*, 131, 769-774.

Vincent-Johnson, N.A., Macintire, D.K., Lindsay, D.S., Lenz, S.D., Baneth, G., Shkap, V., Blagburn, B.L., (1997). A new *Hepatozoon* species from dogs: description of the causative agent of canine hepatozoonosis in North America. *Journal of Parasitology*, 83, 1165-1172.

Vojta, L., Mrljak, V., Curković, S., Zivicnjak, T., Marinculić, A., Beck, R., (2009). Molecular epizootiology of canine hepatozoonosis in Croatia. *International Journal for Parasitology*, 39, 1129-1136.

Wang, C., Ahluwalia, S., Li, Y., Gao, D., Poudel, A., Chowdury, E., Boudreaux, M., Kaltenboek, B., (2010). Frequency and therapy monitoring of canine *Babesia* spp infection by high resolution melting curve quantitative FRET-PCR. *Veterinary Parasitology*, 168, 11-18.

Wasser, S.K., Davenport, B., Ramage, E.R., Hunt, K.E., Parker, M., Clarke, C., Stenhouse, C., (2004). Scat detection dogs in wildlife research and management: application to grizzly and black bears in the Yellowhead Ecosystem, Alberta, Canada. *Canadian Journal of Zoology*, 82, 475–492.

Wlosniewski, A., Leriche, M.A., Chavigny, C., Ulmer, P., Donnay, V., Boulouis, H.J., Mahl, P., Druilhe, P., (1997). Les porteurs asymptomatiques de *Babesia canis* dans une region enzootique. *Comparative Immunology, Microbiology and Infectious Diseases*, 20, 75-86.

Wong, S.S., Teng, J.L., Poon, R.W., Choi, G.K., Chan, K.H., Yeung, M.L., Hui, J.J., Yuen, K.Y., (2011). Comparative evaluation of a point-of-care immunochromatographic test SNAP 4Dx with molecular detection tests for vector-borne canine pathogens in Hong Kong. *Vector Borne Zoonotic Diseases*, 11, 1269-77.

Yamane, I., Thomford, J.W., Gardner, I.A., Dubey, J.P., Levy, M., Conrad, P.A., (1993). Evaluation of the indirect fluorescent antibody test for diagnosis of *Babesia gibsoni* infections in dogs. *American Journal of Veterinary Research*, 54, 1579-1584.

Yamasaki, M., Hwang, S.J., Ohta, H., Yamato, O., Maede, Y., Takiguchi, M., (2008). Flow cytometry to evaluate the level of *Babesia gibsoni* parasitemia *in vivo* and *in vitro* by using the fluorescent nucleic acid stain SYTO16. *Japanese Journal of Veterinary Research*, 55, 129-136.

Yamasaki, M., Kobayashi, Y., Nakamura, K., Sasaki, N., Murakami, M., Rajapakshage, B.K., Ohta, H., Yamato, O., Maede, Y., Takiguchi, M., (2011). *Babesia gibsoni*: detection in blood smears and formalin-fixed, paraffin-embedded tissues using deoxyribonucleic acid in situ hybridization analysis. *Experimental Parasitology*, 127, 119-126.

Zahler, M., Rinder, H., Schein, E., Gothe, R., (2000). Detection of a new pathogenic *Babesia microti*-like species in dogs. *Veterinary Parasitology*, 89, 241-248.

Zhou, J., Fukumoto, S., Jia, H., Yokoyama, N., Zhang, G., Fujisaki, K., Lin, J., Xuan, X., (2006a). Characterization of the *Babesia gibsoni* P18 as a homologue of thrombospondin related adhesive protein. *Molecular and Biochemical Parasitology*, 148, 190-198.

Zhou, J., Yang, J., Zhang, G., Nishikawa, Y., Fujisaki, K., Xuan, X., (2006b). *Babesia gibsoni*: An apical membrane antigen-1 homologue and its antibody response in the infected dogs. *Experimental Parasitology*, 114, 329-333.

Zhou, J., Jia, H., Nishikawa, Y., Fujisaki, K., Xuan, X., (2007). *Babesia gibsoni* rhoptry-associated protein 1 and its potential use as a diagnostic antigen. *Veterinary Parasitology*, 145, 16-20.

Revised by Dr Catalina Rey-Valeirón, Departamento de Sanidad Animal, Programa de Veterinaria, Universidad Nacional Experimental Francisco de Miranda, Coro, Venezuela.

In: Carnivores
Editors: F. I. Álvares and G. E. Mata
ISBN 978-1-62257-323-3

Chapter 2

THE RACCOON DOG (*NYCTEREUTES PROCYONOIDES*) IN THE COMMUNITY OF MEDIUM-SIZED CARNIVORES IN EUROPE: ITS ADAPTATIONS, IMPACT ON NATIVE FAUNA AND MANAGEMENT OF THE POPULATION

***Kaarina Kauhala*[1,*] *and Rafał Kowalczyk*[2]**
[1]Finnish Game and Fisheries Research Institute, Turku, Finland,
[2]Mammal Research Institute, Białowieża, Poland

ABSTRACT

Raccoon dog (*Nyctereutes procyonoides*) is an alien carnivore in Europe. It is today widespread and common in many parts of the continent. Here we review its adaptations to various environmental and climatic conditions, impact on the native fauna and management of the populations. The raccoon dog is very adaptable due to its versatile diet, ability to gather large fat reserves in autumn and to hibernate in areas where winters are harsh. It also has higher than expected reproductive

* Email: kaarina.kauhala@rktl.fi.

potential, which compensates the high mortality rate of juveniles. Raccoon dogs are monogamous and the male participates in pup rearing. Since no helpers have been observed, juveniles usually disperse during their first autumn and some of them wander far. In a colonizing population also adults have a tendency to wander long distances. Due to massive and dispersed introductions, raccoon dogs are characterized by high genetic variation.

Raccoon dogs fitted well into the guild of medium-sized carnivores and share their home ranges with badgers (*Meles meles*) and foxes (*Vulpes vulpes*), and especially raccoon dogs and badgers partly use the same habitats, such as deciduous forests and fields. Raccoon dogs often settle in active badger setts to avoid predation and unfavourable weather conditions. Raccoon dogs may partly overlap their diet with that of native carnivores. Raccoon dogs can spread diseases and parasites, such as rabies and sarcoptic mange, to other carnivores due to their frequent opportunities of contact.

The raccoon dog is a true omnivore and eats anything it can find and catch. Although it sometimes preys on birds, there is no firm evidence that it affects bird populations. The occurrence of birds in its diet shows great spatial and seasonal variation but in most areas is fairly low. Raccoon dogs may have, however, an impact on amphibian populations, especially on islands.

Raccoon dogs are actively hunted in many countries. In Finland, the annual hunting bag is around 160,000 individuals, i.e. about 50% of the autumn population. In spite of many campaigns to reduce the population, the annual hunting bag has steadily increased during the past two decades in Finland. The hunting bag has also increased exponentially in Germany since the early 1990s. Due to their adaptability, large distribution and secretive life, raccoon dogs are very difficult to eradicate after they have established in an area.

INTRODUCTION

The raccoon dog (*Nyctereutes procyonoides*), which originates in eastern Asia, was introduced by Russians mainly to the European parts of the former Soviet Union during the first half of the 20th century (Lavrov 1971, Helle and Kauhala 1991). Due to its adaptability, tendency to wander and high reproductive potential it expanded its range rapidly to many eastern and northern European countries. Today it is among the most numerous carnivores in many countries and it is still expanding its range westwards and southwards in Europe.

Being an alien carnivore, the raccoon dog has been accused to cause damage especially to native game bird populations. It has also been suspected to compete with native carnivores for the best habitats or food resources. The raccoon dog has also been recognized as an important vector of diseases and parasites in its new range. Especially hunters' organizations have claimed intensive control or even eradication of raccoon dog populations but, despite of several campaigns to reduce its numbers, its populations seem to be increasing in Europe.

In this chapter we review the raccoon dog's adaptability, i.e. the reasons behind its success in Europe, its impact on native fauna, its role as a vector of diseases and parasites and the possibility to control its populations.

1. An Adaptable Carnivore

1.1. Omnivory and Diet Variation

Although the raccoon dog is a member of Carnivora, its diet is highly versatile and it is a true omnivore. This can even be seen in its morphology (Bannikov 1964): it has a long gut and small skull and tiny canine and carnassial teeth compared e.g. to those of the red fox (*Vulpes vulpes*), another canid of similar size. Its molars, adapted to grind plant and insect material, are well developed (Kauhala et al. 1998a). Raccoon dog's teeth thus reflect its omnivorous diet, which varies according to the availability of different food items and thus shows spatial and seasonal variation (e.g. Nasimovič and Isakov 1985, Sutor et al. 2010). Accordingly, the food niche of the raccoon dog is much wider than that of other medium-sized carnivores (Jędrzejewska and Jędrzejewski 1998, Kauhala et al. 1998b).

In many areas of Europe small mammals constitute a major part of the raccoon dog's diet (Kolosov et al. 1965, Kauhala et al. 1993a, 1998b, Sutor et al. 2010; Figure 1). Besides small rodents, raccoon dogs commonly prey on shrews (*Sorex* spp): 13-23% of raccoon dog scats contained remains of shrews in summer in southern Finland (Kauhala and Auniola 2001). Larger mammals, such as hares (*Lepus* spp.) occur occasionally in the diet but were probably consumed as carrion.

The occurrence of birds in the raccoon dog's diet varies according to season and area and increases with latitude (Sutor et al. 2010). For instance, in an island in SW Finland, birds occurred in 61% of raccoon dog scats (Ihalainen 2012), whereas in Germany remains of wild birds and eggs were found only in 7.5% of raccoon dogs' stomachs in summer (Sutor et al. 2010). Birds occur more often in the diet in years when voles are scarce than during years when voles are abundant (Ivanova 1962, Judin 1977, Kobylińska 1996). Many of the larger birds (e.g. ducks) were probably eaten as carrion (Barbu 1972, Kauhala and Auniola 2001).

Carrion occurs in the diet especially in winter in areas (e.g. Belarus, Poland and Germany) where raccoon dogs do not hibernate (Jędrzejewska and Jędrzejewski 1998, Sidorovich et al. 2000, Sutor et al. 2010). Ungulate carcasses are an important food source for raccoon dogs in winter and may affect their survival during the harshest period of the year. Carrion and other waste (e.g. human refuse from compost heaps) occurred also in about 50% of stomachs in all seasons in Finland (Kauhala et al. 1993a).

Amphibians were common in the diet of raccoon dogs especially in wetland areas where the share of small mammals was smaller (Barbu 1972, Sutor et al. 2010). Even frogspawn was consumed by raccoon dogs (Ihalainen 2012). Frogs and reptiles were more common in the diet on Finnish mainland, where many vipers (*Vipera berus*) are killed by cars, than on islands where no roads exist (e.g. Kauhala et al. 1998b).

Fish occurred in the diet especially in Finland in late winter when raccoon dogs pick small fish from the ice where fishermen discarded them (ice-fishing is popular in Finland; Kauhala et al. 1993a). Fish were also eaten by raccoon dogs in a wetland area in Germany where 11% of stomachs contained remains of fish (Sutor et al. 2010).

Insects and other invertebrates are commonly found in raccoon dogs' diet in all areas (Sutor et al. 2010). In summer, almost every scat contains remains of invertebrates, especially Coleoptera and Hymenoptera (Kauhala et al. 1993a). The most common beetles in the diet were Carabidae, Geotrupidae and Silphidae (Ihalainen 2012). Due to the small size of invertebrates the frequency of occurrence may, however, exaggerate their importance for raccoon dogs.

Raccoon dogs consumed berries and fruits frequently all over Europe (Sutor et al. 2010). Fruits and berries, such as bilberries (*Vaccinium myrtillus*) and lingonberries (*V. vitis-idaea*), are important food sources in late summer and autumn when raccoon dogs are accumulating fat reserves for winter (Nasimovič and Isakov 1985, Kauhala et al. 1993a).

Raccoon dogs frequently visit gardens where they consume fruits, such as *Prunus* spp., *Malus* spp., *Pyrus* spp. and *Vitis* spp. (Nasimovič and Isakov 1985, Kauhala 2009, Ihalainen 2012).

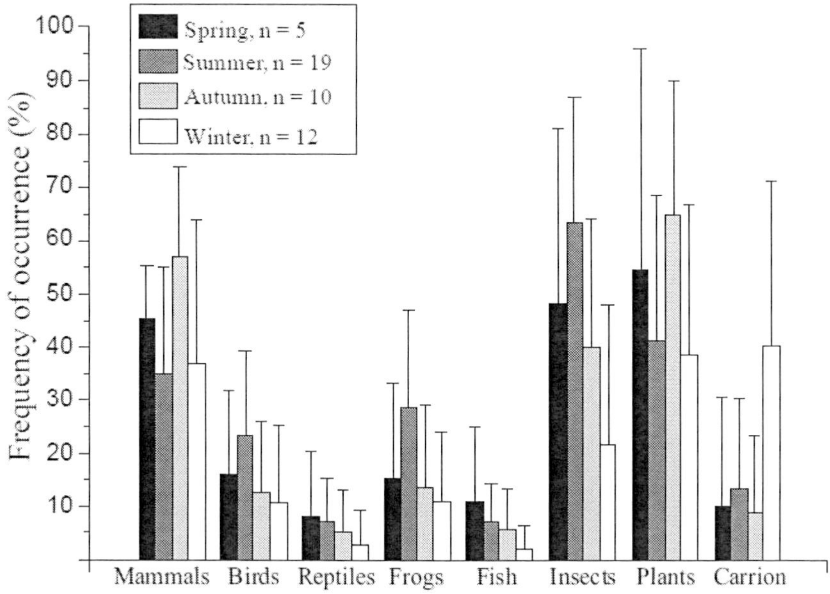

Figure 1. The diet of raccoon dogs in Europe in different seasons (pooled data from several areas). N indicates the number of data sets. Frequency of occurrence gives the percentage stomachs/feces which contained each food item. 'Insects' may contain some other invertebrates. Original figure in Kauhala 2009.

The raccoon dog's versatile diet is one important factor behind its success in Europe: raccoon dogs can always find something to eat despite of habitat and season. The exception is harsh winters with thick and soft snow cover. In these conditions raccoon dogs are not able to move around and find food. They have solved the problem by gathering large fat reserves in autumn and hibernating in areas with harsh winters.

1.2. Hibernation and Fat Reserves

The raccoon dog is the only canid that hibernates in winter in northern areas. In southern Finland, hibernation usually starts in November and ends in March, i.e. during the season when food resources are sparse and difficult to find (Kauhala et al. 2007). Most raccoon dogs sleep when the air temperature is below -10 °C, the snow depth is > 35 cm and day length is < 7 hours, and most are active when the ambient temperature is above zero, there is no snow and the day length is > 10 hours (Kauhala et al. 2007).

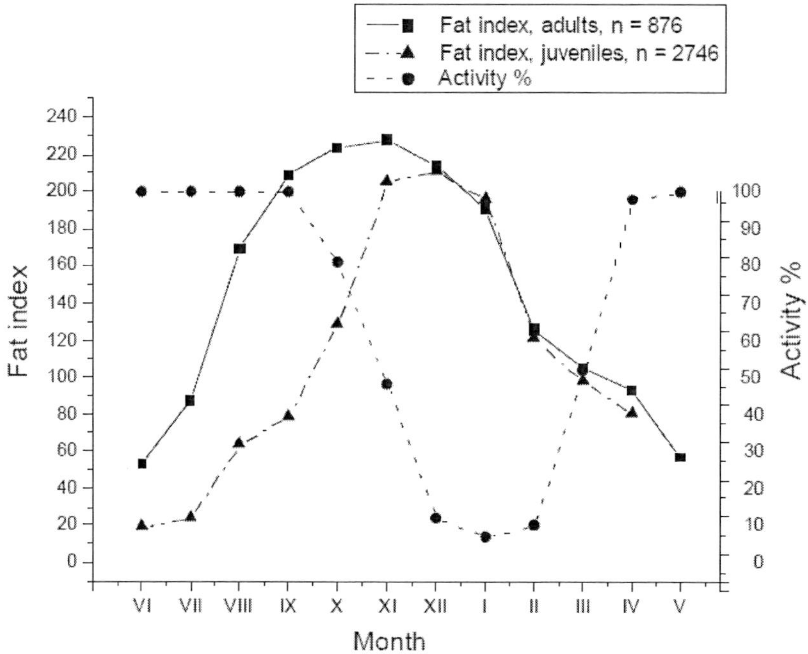

Figure 2. The mean fat indices of adult and juvenile raccoon dogs and activity % in different months. Fat index = 100* (amount of fat, g, around kidneys and at the top of the abdominal cavity)/head-body length (cm). Data are based on raccoon dog carcasses collected from hunters in Finland between 1986 and 1999. N = number of individuals examined. Activity % = percent raccoon dogs active in a given month, according to radio-tracking in SE Finland between 2000 and 2003.

Raccoon dogs often wake up temporarily during winter and change the dens a few times. Raccoon dogs do not hibernate in more southern areas, e.g. in Germany (Drygala et al. 2008a). When hibernating, they settle in good shelters such as badger setts or hollow trees (Kauhala et al. 2007, Kowalczyk et al. 2008, Kowalczyk and Zalewski 2011).

Raccoon dogs accumulate large fat reserves and almost double their weight between early summer and late autumn (Kauhala 1993). Adults fatten themselves earlier than juveniles and reach their maximum mass and fat reserves in October-November, whereas juveniles reach their maximum mass one or two months later (Figure 2). Juveniles reach the adult body length in late autumn, usually in November (Kauhala 1993). Raccoon dogs lose about 43% of their body mass during hibernation (Mustonen et al. 2007). Fat accumulation in autumn and hibernation are regulated by hormonal chances, especially by thyroid hormone level, which is low during hibernation (Korhonen 1987, Nieminen et al. 2002).

The average body temperature of raccoon dogs is 38 °C in the snow-free season (Mustonen et al. 2007). During hibernation their body temperature decreases 1.4 – 2.1°C. Circadian variation in body temperature is greater in winter than in summer and body temperature is lowest in the morning (06:00-11:00 a.m.) and highest in the evening (16:00-23:00 p.m.) (Mustonen et al. 2007). In winter, the circadian activity of raccoon dogs is also lowest in the morning and highest around midnight (Kauhala et al. 2007).

In areas where summers are long and winters short juveniles have enough time to fatten themselves before winter lethargy. Further north, such as northern Finland, summers are too short and winters too long for juveniles to accumulate enough fat and survive the winter (Kauhala 1993; Kauhala and Helle 1995). The length of the summer thus determines the northern limit of permanent raccoon dog distribution in Europe, which today lies near the Arctic Circle. However, without winter sleep raccoon dogs could not live so far north they do nowadays, because it would be impossible for them to move in deep snow and to find food in winter.

1.3. High Reproductive Potential

Among canids, raccoon dogs have exceptionally large litters relative to their body weight (Kauhala 1992). In favourable areas, both in the introduced and original ranges, the mean litter size is 8-10 pups (Judin 1977, Helle and Kauhala 1995, Kowalczyk et al. 2009). Maximum recorded litter size in Finland was 16 pups, and maximum number of *corpora lutea* was 23 (Helle and Kauhala 1995). Omnivory, monogamy and winter sleep are probable factors behind the high reproductive potential of raccoon dogs (Kauhala 1996a): some food is always available, males help in pup-rearing and due to winter sleep females are in fairly good condition at the time of ovulation despite of the weather conditions in winter.

Raccoon dogs reach sexual maturity at the age of 10 months (Helle and Kauhala 1995, Kowalczyk et al. 2009). The following figures are based on carcass material of 599 mature females examined in Finland (Helle and Kauhala 1995). Sixty-six percent of one-year old individuals give birth, and the proportion of reproducing females is higher in older females (78% at 2 years, 88% at 3 years, 84% at 4 years and 93% thereafter). The litter size, on the basis of both placental scars and embryos/fetuses, of 1-2-year-old females is lower (mean 8.7) than that of middle-aged (3-5-year old) females (mean 9.5). The litter size declines thereafter, being 8.5 among 6-7-year old females. Although productivity of 1-year-old females is lowest (5.7 pups/female), they produce a large proportion of pups in the population due to their high numbers. The reproductive value of females is highest at the age of two years due to the high mortality rate in the younger age classes (Helle and Kauhala 1993).

Climate affects the productivity of a raccoon dog population. The time of ovulation depends on weather conditions in spring: when snow melts early, pups are born earlier and have more time to grow and accumulate fat reserves before winter than juveniles in the more northern areas (Helle and Kauhala 1995, Kauhala and Helle 1995, Melis et al. 2010). In areas with long winters, raccoon dog females have less fat, produce smaller litters, and a smaller proportion of them reproduce than females in areas with short winters and long summers (Helle and Kauhala 1995; Figure 3). Also food resources affect pup production. Abundance of small rodents affects the fat reserves of raccoon dogs and thus also their litter size (Kauhala and Helle 1995). The crop of berries may even affect the yearly variation in raccoon dog numbers, as it affects the mortality rate of fetuses and probably also that of juveniles during their first autumn and winter (Helle and Kauhala 1991, Helle and Kauhala 1995).

1.4. Monogamy and Parental Care

Radio-tracking studies have shown that raccoon dogs usually are monogamous (e.g. Kauhala et al. 1993b, Drygala et al. 2008a, 2008b). Male and female have a permanent pair-bond which usually lasts for their lifetime. They share their home range and especially in the pup-rearing season, home ranges of neighbouring pairs do not usually overlap. Male and female move and rest together the year round, except in the pup-rearing season in May-June when males spend more time than females at the den guarding and warming the pups (Kauhala et al. 1993b, Kauhala et al. 1998c, Drygala et al. 2008b, Kowalczyk and Zalewski 2011). At daytime males are usually 'baby-sitting', whereas females forage. When females nurse the pups they need a lot of energy and spend more time foraging than males. At night they take turns: one is with pups while the other is foraging. During the first month after birth pups are seldom left alone (Ikeda 1983, Yamamoto 1987, Kauhala et al. 1998c).

When pups are 6-8 weeks old, they start to move around the den site. They may follow their parents at night and start foraging themselves. They catch small food items, such as insects, small birds (probably young individuals) and eggs (Kauhala et al. 1998b). In late summer juveniles start to move on their own and most of them disperse during their first autumn (Kauhala et al. 1993b).

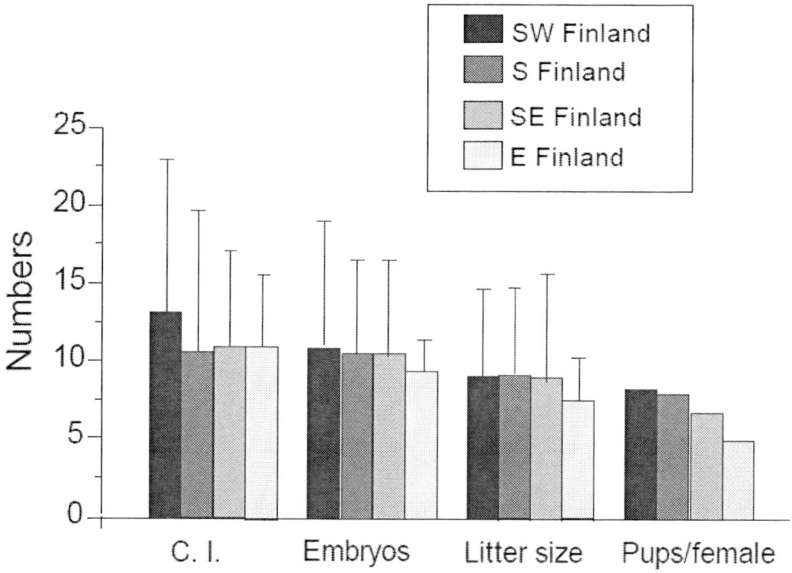

Figure 3. The mean numbers (± SD) of corpora lutea (C. l.), embryos, pups at birth (litter size) and productivity (pups/adult female) of raccoon dog populations in different areas of Finland. Climate is mildest in SW Finland and coldest in E Finland. The mean temperature of the year is approximately 4.7 °C and the length of winter about 130 days in SW Finland, the corresponding figures for the other areas being: 3.8 °C and 140 days for S and SE Finland and 2.4°C and 175 days for E Finland. Snow depth at the end of March is usually zero in SW Finland, about 5 cm in S Finland, 35 cm in SE Finland and 65 cm in E Finland (Kauhala and Helle 1995, the Finnish Meteorological Institute).

1.5. Tendency to Wander

The mean dispersal distances of juvenile raccoon dogs in southern Finland were estimated to 14-19 km but some individuals may wander as far as 50-70 km, even 145 km, from the marking place (Kauhala et al. 1993b, Kauhala and Helle 1994, Kauhala et al. 2006). Fifty percent of recoveries of ear-tagged juveniles were made within 5 km from the marking place but 21% wandered more than 20 km and 9% more than 40 km (Kauhala et al. 1993b). In northern Germany, the mean dispersal distances of juveniles of both sexes were 13.5 km, and 56% of juveniles were recovered within 5 km from the marking place (Drygala et al. 2010).

The maximum distance recorded in northern Germany was 91 km. Both in Finland and in northern Germany most juveniles disperse in September and October but some may disperse as early as in July, and some may not leave their natal home range before late winter (Kauhala et al. 1993b, Drygala et al. 2010). Dispersal is equal between sexes, probably due to monogamy of the species and the fact that no helpers have been found in raccoon dogs.

On a more or less stable population only young individuals disperse, but adults stay in their home ranges from year to year (Nasimovič and Isakov 1985, Kauhala et al. 1993b, Kauhala and Helle 1994). In a colonizing population also adults may disperse, and dispersal distances are sometimes several hundreds of kilometers (Nasimovič and Isakov 1985, Sutor 2008). Nowak (1973) reported of a dispersal distance of 500 km within three years in a colonizing population. In Sweden, an adult male with a GPS-collar wandered 650 km in a few months (Mikkola 2011).

1.6. Genetic Variation

Genetic variability of the raccoon dog in Europe is probably an effect of combination of different factors including large scale mass introduction, multiple translocations, use of large number of individuals from different geographical areas, and admixture of different mtDNA lineages (Kauhala and Kowalczyk 2011). Phylogenetic analysis indicates different invasion corridors of the species in central Europe (Ansorge et al. 2009) with the secondary contact zone in Germany between different source populations (Pitra et al. 2010).

During expansion, some populations were probably created from small number of founders, so population went through several bottlenecks (especially during initial introductions) followed by hybridization of individuals from divergent haplogroups originating from multiple source populations from the native range.

Nine haplotypes were found in raccoon dogs in Europe with a sequence divergence of 0.2%–3.2% (mean 1.3%). Haplotype diversity (*h*) was quite high and varied from 0.676-0.746 in populations in Germany to 0.837 in Finland (Pitra et al. 2010).

2. A Vacant Niche? Relationship with Other Medium-sized Carnivores

2.1. Habitat Use, Home Ranges and Densities

Raccoon dogs prefer damp meadows, marshland and open woodlands where the undergrowth is abundant offering shelter and food for raccoon dogs (Judin 1977, Nasimovič and Isakov 1985, Kauhala and Auttila 2010, Kauhala et al. 2010). They also favour deciduous and mixed forests and fields and often move along lake and river shores (Kauhala 1996b, Holmala and Kauhala 2009). They even escape into water when chased e.g. by a dog (pers. obs.). Raccoon dogs avoid large coniferous forests and other forests with thick canopy and sparse undergrowth (Nasimovič and Isakov 1985, Kauhala and Auttila 2010). However, habitat preferences differ between areas and in northern Germany, no habitat preferences were found (Drygala et al. 2008c).

In many areas they invaded, raccoon dogs became more numerous than native medium-sized carnivores – badgers and red foxes (Kowalczyk et al. 2009). In Finland, the population density of raccoon dogs is highest (up to 2 individuals per 100 ha) in areas which consist of a small scale mosaic of meadows, small deciduous/mixed forests, fields and gardens (Kauhala et al. 2010). Accordingly, home range size of raccoon dogs shows spatial variation. Home ranges (kernel 95%) are smallest (seasonal home ranges about 100 ha) in the most suitable areas but much larger (mean 260 ha in summer) in areas where coniferous forests are common, such as SE Finland (Kauhala et al. 2010). Summer home ranges in northern Germany are almost as large as those in SE Finland (mean males: 194 ha, mean females: 225 ha; Drygala et al. 2008a). In continuous woodlands of the Białowieża Primeval Forest, raccoon dog densities are 0.5-0.7 ind/km^2 (Jędrzejewska and Jędrzejewski 1998).

Raccoon dogs and badgers (*Meles meles*) often share their home ranges but have partly different habitat preferences (Kauhala and Auttila 2010). Although their home ranges largely overlap, the core areas of their ranges tend to be mutually exclusive or overlap only slightly in areas where the most preferred habitats for each species exist (Figure 4). When raccoon dogs prefer damp meadows with abundant undergrowth, badgers favour mature deciduous forests with sparse undergrowth (Kauhala and Auttila 2010). However, when the most preferred habitats for each species do not exist in the area, both species use similar habitats, such as young mixed or deciduous forests and fields (Holmala and Kauhala 2009, Kauhala and Auttila 2010). Both species are thus very flexible in their habitat use, which facilitates the coexistence of these two species. A radio-tracking study in SE Finland showed, however, that badgers used the preferred habitat in the area (deciduous forest) more often before midnight and raccoon dogs more often after midnight (Kauhala and Holmala 2008). The temporal segregation of habitat use may indicate the need to avoid direct competition between these species.

Raccoon dogs defecate in a dung pile or latrine, which several individuals may use and which may be used for several years (Ikeda 1984, Yamamoto 1984, Kauhala and Salonen 2012). Many latrines are in the core areas of home ranges but some also lie along the range borders. Latrines are usually located in forests, although raccoon dogs often move also in meadows and field edges (Kauhala and Salonen 2012).

Also badgers defecate in certain latrine sites, each of which usually consists of several small pits. In an island in SW Finland, almost every second badger latrine was within a few meters from a raccoon dog latrine (Kauhala and Salonen 2012). Most of these latrines were in oak-hazel groves, which is the preferred habitat for badgers in the area (Kauhala and Auttila 2010).

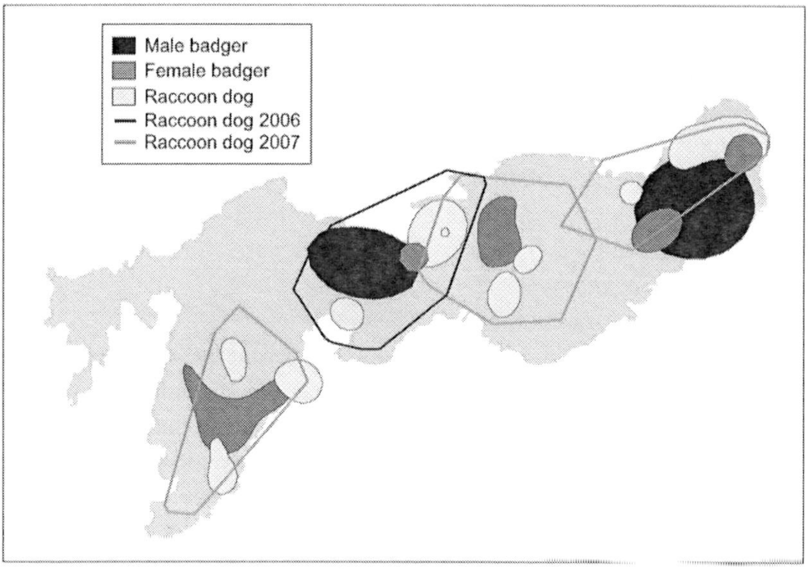

Figure 4. Core areas (kernel 50% home ranges) of raccoon dogs and badgers, and raccoon dog home ranges in an island in SW Finland. Raccoon dogs and badgers were radio-tracked in summers of 2006 and 2007.

We do not know which species made its latrine in the site first. Badger latrines were, however, situated more often in the core areas of raccoon dogs than *vice versa*. Furthermore, raccoon dog latrines were in permanent places, which may indicate that raccoon dogs were there first and badgers then defecated nearby. These observations point to the conclusion that intra-guild competition for space may exist between raccoon dogs and badgers.

In Finland, the most preferred habitats (mature oak-hazel groves) for badgers are very rare and badger density accordingly much lower than that of raccoon dogs (Kauhala et al. 2006). Badger density is highest in areas with a high proportion of mixed forests and low in areas with vast spruce forests (Kauhala and Holmala 2011).

Fox home ranges are usually larger than those of raccoon dogs, e.g. in SE Finland fox home ranges averaged 660 ha whereas raccoon dog home ranges averaged 390 ha (Kauhala et al. 2006). Fox density in the area was about half (3.5 individuals/10 km^2) of the estimated raccoon dog density (7.7 individuals /10 km^2). Habitat use of raccoon dogs and foxes partly overlapped: both species selected young mixed forests from the landscape (Holmala and Kauhala 2009). Raccoon dogs favoured, however, more often fields than did foxes, whereas foxes preferred more often coniferous forests.

There are very little data of the possible competition between raccoon dogs and foxes. Data collected during a predator removal study in southern Finland suggested, however, that when raccoon dogs were removed from the study area, fox population (snow track index) started to increase (Kauhala 2004). Furthermore, when feces of medium-sized carnivores were collected in SW archipelago of Finland, signs of raccoon dogs and foxes simultaneously inhabiting small islands were not found: in small islands there were either signs of foxes or raccoon dogs (Kauhala and Auniola 2001).

2.2. Shelter Use

Raccoon dogs almost always hide when resting. Most often they rest in dense vegetation (sedges, thickets), under natural shelters created by branches, roots, and trunks of fallen trees, and in different burrows and in hollow trees. Patterns of shelter use may vary according to season, temperature or forest structure (Kowalczyk and Zalewski 2011). To avoid predation and harsh weather conditions raccoon dogs settle in places offering good shelter against predators and harsh weather conditions (Kowalczyk and Zalewski 2011). They hibernate and rear their pups in dens or other shelters. The same dens are often used for several years by the same pairs of raccoon dogs for wintering or reproduction (Kowalczyk and Zalewski 2011). In SE Finland, raccoon dogs used, however, the mean of 3.3 dens during winter (Kauhala et al. 2007). Raccoon dogs may also hide their pups in large dry reed beds (pers. obs.). Reed beds are warmer and there are probably fewer parasites than in burrows.

Raccoon dogs often use badger setts occupying them simultaneously with badgers. Badgers usually start hibernation earlier in autumn and wake up later in spring than raccoon dogs and thus may not be aware of the fact that raccoon dogs accompany them in the den. Sometimes several badgers and raccoon dogs hibernate in the same den and occasionally share the den also in summer (Kauhala and Holmala 2006), but they usually utilize different parts of setts. However, aggressive interactions among the two species were observed, including killing of raccoon dog pups by badgers (Kowalczyk et al. 2008). Because raccoon dogs are not so well adapted to dig dens as are badgers, the occurrence of badgers and their setts in the area may play a facilitative role and help raccoon dogs to colonize the area (Kowalczyk et al. 2008).

2.3. Overlap of Food Niches between Medium-sized Carnivores

Raccoon dogs, badgers and foxes are all omnivores with a versatile diet. Foxes tend to be, however, more carnivorous than the other two, at least in northern latitudes (Kauhala et al. 1998b). Major part of the diet of foxes in Finland consisted of mammals and birds, but foxes also consumed invertebrates, berries and human refuse (Kauhala et al. 1998b). The food niche of raccoon dogs was wider than that of foxes or badgers consisting of small rodents and shrews, some birds, frogs and lizards, invertebrates, fruits and berries, carrion and waste (Kauhala et al. 1998b, Sutor et al. 2010). The badger also had a versatile diet but consumed more frogs and invertebrates than raccoon dogs (Kauhala et al. 1998b).

In Białowieża Primeval Forest (Poland), food niche overlap between native badgers and red foxes in spring and summer was very low (2%). Food niche overlap was much higher between badgers and raccoon dogs (35%) and highest between foxes and raccoon dogs (41%). In winter, when both foxes and raccoon dogs feed more often on carrion, overlap of their food niche increased to 62% (Jędrzejewska and Jędrzejewski 1998).

The flexible diet of all three species diminishes the possibility of severe competition between them and it seems that they can share the food resources, at least in southern Finland. Furthermore, hibernation of raccoon dogs and badgers in northern areas, such as Finland, results in a situation where there is no competition for food in winter, which is the harshest period of the year. As very efficient scavengers, raccoon dogs may, however, compete with other carnivores for ungulate carcasses (Jędrzejewska and Jędrzejewski 1998, Sidorovich et al. 2000). Sidorovich et al. (2000) pointed out that other carnivores suffer from competition for carcasses in early spring, when other resources are scarce, and their populations declined in Belarus when raccoon dog numbers increased. The polecat (*Mustela putorius*) is one of the species which may have suffered most from the competition for carcasses in winter.

The dentition of Finnish red fox females has slightly changed after the arrival of raccoon dogs in Finland (Viranta and Kauhala 2011). Today the teeth of female foxes show more carnivorous features than before the invasion of the raccoon dog. Vixens are smaller than male foxes, and are thus more likely than males to compete for food with raccoon dogs. This may thus be a case of character displacement in order to avoid competition but also other factors may be involved.

2.4. Changes in Abundance

Estimating raccoon dog population numbers is difficult, because e.g. snow track counts do not give reliable estimates of raccoon dog density due to their winter sleep. We can, however, assume that the change in the hunting bag partly reflects the trend in population numbers. In Finland, hunting bags of the potential competitors of raccoon dogs, i.e. the red fox, badger and also pine marten (*Martes martes*) have not changed much during the past 15 years, possibly indicating that their numbers have not declined, although raccoon dog catch has increased almost three-fold (Figure 5). In Białowieża Primeval Forest, badger densities have not changed after raccoon dog expansion (Kowalczyk et al. 2000). These facts point to the conclusion that there was a vacant niche for the raccoon dog in most of invaded areas. In Belarus, raccoon dogs may have caused a decline of other predators and raptors because of competition for carrion in early spring, when other food resources are scarce (above; Sidorovich et al. 2000).

Although no firm evidence of competition between raccoon dogs and foxes exists, in some areas of Finland, fox numbers tend to decline when raccoon dog population increases rapidly. Furthermore, the fox population is most numerous in SW Finland, whereas the raccoon dog population is most abundant in S and SE Finland (Kauhala 2007).

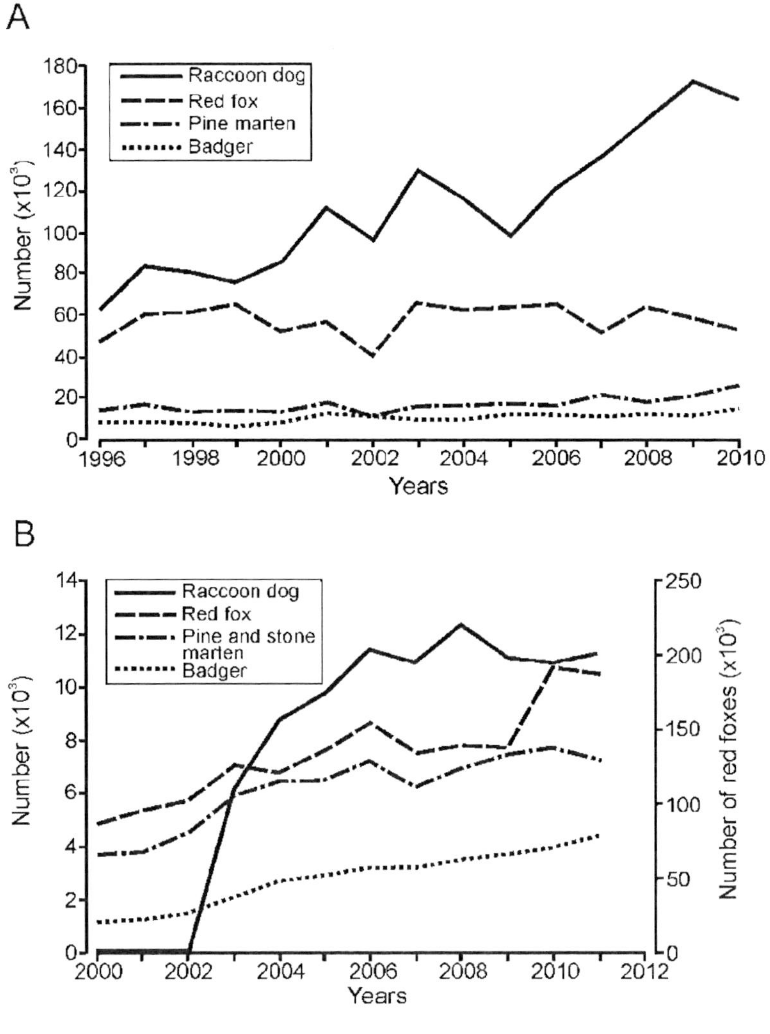

Figure 5. Hunting bags of the raccoon dog, red fox, badger and pine marten in Finland during the past 15 years (Finnish Game and Fisheries Research Institute, A) and in Poland since 2000 (Research Station of Polish Hunting Society, B). Note that the scales are different in the two figures.

2.5. A Vector of Diseases and Parasites

Diseases are an important mortality factor of raccoon dogs (Kowalczyk et al. 2009, Kauhala and Kowalczyk 2011). The raccoon dog's role as a vector of wildlife rabies increased in north-eastern Europe, especially in the Baltic States, during the past few decades (Holmala and Kauhala 2006, WHO 2009). In Europe, there were 148 known cases in raccoon dogs in 1990, whereas the number of known cases was as high as 873 in 2002 (WHO 2002). At the same time period, the number of fox cases decreased from 12,833 to 4,795. In Estonia, there were approximately similar numbers of rabies cases in raccoon dogs and foxes in 2003 but thereafter the number of raccoon dog cases has been higher than that of fox cases (WHO 2003, 2005). Rabid raccoon dogs have also recently been found in Russia, Belarus, Ukraine and Poland (Siemionek et al. 2007, WHO 2010). Furthermore, the raccoon dog was the main vector species (73% of the known cases) during wildlife rabies epizootics in the late 1980s in Finland (Westerling 1991). In Europe, where the fox was the main vector for decades, there is thus a new situation with two important rabies vectors: the raccoon dog and the red fox.

Bait vaccinations have, however, improved the situation, and many European countries are today rabies-free (WHO 2010). For instance, the number of cases has declined sharply in Estonia and Poland during recent years (Figure 6). However, wildlife rabies cases still occur in foxes and raccoon dogs in eastern European countries, most observed cases being in Belarus (544), Croatia (580, not raccoon dogs), Russia (1192) and Ukraine (621; WHO 2010). The total number of rabies cases in wildlife (excluding bats) was 4,310 in Europe in 2010.

Because raccoon dogs share their home ranges and den sites with badgers and foxes and their nightly paths often cross (Kauhala and Holmala 2006), raccoon dogs easily transmit diseases to badgers and foxes. Rabies models constructed for southern Finland showed that an epizootic with two vector species (raccoon dog and red fox) would be stronger than that with only one vector species (Singer et al. 2009). The raccoon dog would be the major host in an area where its density is higher than that of the fox, such as Finland, and the number of fox cases would rather depend on raccoon dog than on fox density. An outbreak of rabies is likely even when the pooled density of the two vector species is lower than the density of each of these species demanded for epizootics with only one vector species (Singer et al. 2009). When the raccoon dog is the main vector species, the risk for an epizootic is highest in autumn when juveniles disperse and in spring when adults move the longest distances (Kauhala et al. 2007). The risk is lowest in winter when raccoon dogs hibernate. Raccoon dogs may also spread dangerous parasites, such as *Echinococcus multilocularis* (Thiess et al. 2001, Machnicka-Rowińska et al. 2002).

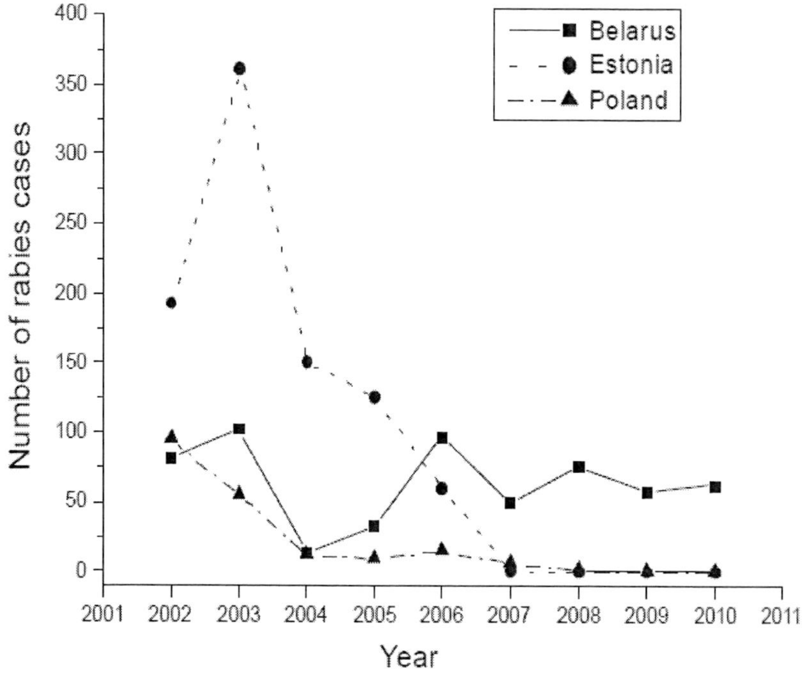

Figure 6. Number of observed rabies cases in raccoon dogs in Belarus, Estonia and Poland in 2002-2010 (WHO).

The fox has been the major definite host of this parasite (small rodents being the intermediate hosts) in Europe but recently its prevalence in raccoon dogs has increased and is now as high as 6-12% in northern Brandenburg (Germany) and 8% in Poland (Machnicka-Rowińska et al. 2002, Schwarz et al. 2011). The parasite is found in many west and central European countries and is spreading northwards in Europe. It reached Estonia by 2005 and Sweden in 2011 (Moks et al. 2005, Romig et al. 2006, Osterman Lind et al. 2011). The first cases in Estonia and Sweden were found from foxes. It is, however, probable that the parasite came to Sweden with dogs brought from some other country (Osterman Lind et al. 2011).

Echinococcus multilocularis is not dangerous to other canids but it can cause a fatal disease (alveolar echinococcosis) in humans. Raccoon dogs also spread, however, parasites dangerous to other wild carnivores. One of these is *Sarpoctes scabiei* causing sarcoptic mange. This painful skin disease can be an important mortality factor among many carnivores, especially among foxes (Shibata and Kawamichi 1999, Kowalczyk et al. 2009, Kauhala and

Kowalczyk 2011). Raccoon dogs are also important reservoirs of *Trichinella* spp. in Finland (Oivanen et al. 2002, Kauhala and Kowalczyk 2011). It carries all fours species of *Trichinella*, which occur in Finland (Oivanen et al. 2002). Furthermore, the prevalence of thrichinosis in wild animals increased in Finland after the arrival of the raccoon dog.

3. Predation on Native Fauna

Raccoon dogs are suspected to cause much damage to game birds, such as waterfowl and grouse populations, by destroying their nests and preying on juvenile and even on adult birds. However, diet studies do not give much support to this (reviews in Judin 1977, Nasimovič and Isakov 1985, Sutor et al. 2010). Most birds consumed by raccoon dogs are passerines (e.g. Kauhala et al. 1993a, Drygala et al. 2000). In rich habitats of the Białowieża Primeval Forest, birds (mainly jay *Garrulus glandarius*) constituted 13% of the biomass consumed in spring-summer, while only 1.5% in autumn-winter. Grouse and waterfowl occur in most areas only in 0-5% of raccoon dog stomachs/feces (e.g. Naaber 1974, Judin 1977, Danilov et al. 1979, Kauhala et al. 1993a, Ansorge 1998, Kauhala et al. 1998b, Sutor et al. 2010). However, in bird colonies or islands raccoon dogs may prey on birds more often than in mainland areas (Naaber 1971, 1984, Ihalainen 2012).

In SW archipelago of Finland, 67% of raccoon dog feces contained remains of ducks, mainly female eiders (*Somateria mollissima*) but it is likely that the birds had died because of a viral disease, which killed many brooding eiders in the area during the years when the samples were collected (Kauhala and Auniola 2001). Raccoon dogs are known to catch sick or injured birds or consume bird carcasses in other areas, too (e.g. Barbu 1972, Naaber 1974, Viro and Mikkola 1981).

Predator removal studies conducted in Finland do not give much support to the harmfulness of raccoon dogs for game bird populations (Kauhala et al. 2000, Kauhala 2004). In southern Finland, breeding success of ducks correlated positively with raccoon dog abundance both in a predator protection area and a predator removal area where most removed predators were raccoon dogs (Kauhala 2004). In contrast, chick production tended to decrease in the predator removal area at the end of the experiment, together with an increasing fox population. The breeding success of grouse did not correlate with raccoon dog abundance in the study areas, except that of black grouse (*Tetrao tetrix*), which correlated positively with raccoon dog abundance in the predator protection area (Kauhala et al. 2000).

Some pieces of egg shells, including those of domestic poultry, are found in raccoon dog feces but they are usually difficult to identify (Kauhala 2009). In southern Finland, remains of egg shells were found in 11-40% of scats (Kauhala and Auniola 2001). These include egg shells from compost heaps. During a nest predation experiment Opermanis et al. (2001) found that raccoon dogs destroyed only 0.3% of waterfowl nests in a wetland area in Latvia. Predator removal or nest predation studies do thus not give firm evidence that raccoon dogs are harmful to game birds.

Raccoon dogs also commonly prey on frogs (e.g. Barbu 1972, Kauhala and Auniola 2001, Sutor et al. 2010). In SW Finland, frogs were, however, almost absent in the raccoon dog's diet in small islands, which may point to the conclusion that frog populations on the islands are sparse. Frogs were easy prey for alien predators, i.e. raccoon dogs and American minks (*Neovison vison*), and frog populations may have already been destroyed by these.

4. Management of the Population

4.1. Expansion and Distribution

The native distribution range of raccoon dogs in the Far East covers northern Vietnam, Korea, Japan, Amur and Ussuri regions of Siberia and eastern China (Kauhala and Saeki 2004). After mass introduction (over 9,000 individuals released in total) to the European parts of the former Soviet Union between 1929 and 1955, the raccoon dog has spread over eastern, northern and central Europe (Nowak and Pielowski 1964). Pace of spread was around 40 km per year (Lavrov 1971). By the 1980s raccoon dogs had colonized an area over 1.4 million km^2 of Europe (Nowak and Pielowski 1964, Nowak 1973), and still continues its expansion. It reached Finland in 1930s, Sweden in 1945, Baltic countries and Romania in 1950s, Poland in 1955, eastern Germany, Austria and Hungary in early 1960s (Nowak and Pielowski 1964, Kauhala and Kowalczyk 2011).

After first phase of fast expansion and rapid population growth, raccoon dog invasion slowed down. The factors behind this were probably man-changed habitats in western Europe, less favourable for the species strongly linked to wet natural habitats. In late 1970s raccoon dogs reached France, in 1980s Balkan countries and Norway. In 2000s, raccoon dogs have managed to cross the Alps, and were recorded in northern Italy (P. Genovesi, in litt.). Nowadays, raccoon dogs are found occasionally in Bulgaria, the Netherlands, Macedonia, Switzerland, and even in Spain (Kauhala and Kowalczyk 2011).

4.2. Mortality

Raccoon dog populations are characterized by a very high mortality rate. The annual mortality rate of the population in southern Finland was 81%, being 51-54% for adults and 88-89% for juveniles (Helle and Kauhala 1993). Hunting is a major cause of mortality in Finland. In Białowieża Primeval Forest (Poland) the mortality rate of adults, mainly from predation (from wolves and dogs) and diseases, was 62%, while 57% of juveniles died during the first three months of their life (Kowalczyk et al. 2009). The majority of raccoon dogs (98%) died before reaching the third year of their life (in Finland 1% reached the age of 5 years) and the maximum life span is 7-8 years. Life expectancy at birth is 0.7-0.8 years (Helle and Kauhala 1993, Kowalczyk et al. 2009). High moratlity of juveniles due to collisions with cars is recorded during dispersal (August-September).

4.3. Hunting in Different Countries

The hunting season for raccoon dogs is from 1 August to 30 April in Finland. The annual hunting bag has increased rapidly since the mid-1990s, being today about 160,000 each year (Finnish Game and Fisheries Research Institute; Figure 5 above).

About 50% of the autumn population is hunted each year (Kauhala 2007). The size of the autumn population would thus be about 320,000 individuals in Finland, including the juveniles which have survived their first summer. Despite of the high hunting pressure, the hunting bag has increased in recent years. This fact indicates that raccoon dogs are very difficult to eradicate, when the population is established and raccoon dog numbers are high.

The hunting bag has also increased rapidly in Germany since the mid-1990s, being today about 30,000 individuals (Drygala et al. 2008a, 2008b, German Hunter's Union). In Poland, 11,000-12,000 raccoon dogs were hunted annually in 2005-2010 (Data by Research Station of Polish Hunting Society, 2012).

Conclusion

Raccoon dog is one of the invasive carnivores that were introduced to Europe in the 20th century. It has adapted very well to habitats of eastern and northern Europe, and continues its invasion to western and southern parts of the continent (Kauhala and Kowalczyk 2011). Successful expansion of the species was due to a combination of mass introductions into multiple locations, its tendency to wander and the high plasticity and adaptability of raccoon dogs. Its high reproductive capacity enabled fast expansion and population growth, despite of the high mortality rate of raccoon dogs. The high plasticity in adaptation to different climatic and environmental conditions, hibernation and the very wide food niche allowed raccoon dogs' adaptation to local conditions in different areas (Kauhala and Kowalczyk 2011). Finally, low and delayed control efforts in many countries and the secretive way of life were responsible for the unrecorded spread and population establishment in new localities. Impact of raccoon dogs on native fauna is of minor importance than expected. Locally they may, however, be an important threat to populations of waterfowl and amphibians. Probably the most severe consequence of raccoon dog introduction and expansion is its role as a vector and reservoir of diseases and parasites, such as rabies, sarcoptic mange, trichinellosis and *Echinococcus multilocularis* (Westerling1991, Oivanen et al. 2002, Oksanen and Lavikainen 2004, Kauhala and Kowalczyk 2011, Kowalczyk and Zalewski 2011). Raccoon dogs have fitted well in local communities of carnivores and coexist with other medium-sized predators: the red fox and badger. Badgers might facilitate the expansion of raccoon dogs, especially in areas occupied by large predators, through habitat amelioration (setts) and refuge from cold and predation (Kowalczyk et al. 2008). Wide distribution, locally high densities and continuation of expansion make eradication of raccoon dogs in Europe almost impossible. Raccoon dogs may additionally benefit from climate warming and disperse to areas where cold climates and harsh winters limit today their distribution.

References

Ansorge, H. (1998). Biologische Daten des Mardeshundes aus der Oberlausitz. *Abhandlungen und Berichte des Naturkundemuseums Görlitz, Band 70*, 47-61.

Ansorge, H., Ranyuk, M., Kauhala, K., Kowalczyk, R. and Stier, N. (2009). Raccoon dog *Nyctereutes procyonoides* populations in the area of origin and in colonised regions – the epigenetic variability of an immigrant. *Annales Zoologici Fennici 46*, 51-62.

Bannikov, A.G. (1964). [Biology of the raccoon dog in U. S. S. R.] *Mammalia 28*, 1–39 (In French).

Barbu, P. (1972). [Results of studies of raccoon dogs *Nyctereutes procyonoides ussuriensis* Matschie, 1907, from the Danube Delta.] *Säugetierkundliche Mitteilungen 20*, 375-405 (In German).

Danilov, P. I., Rusakov, O. S. and Tumanov, I. P. (1979). [Carnivores of the northwest Soviet Union.] Hiŝnye zveri severo-zabada SSSR. *Nauka, Leningrad*. Pp. 5-25 (In Russian).

Drygala, F., Mix, H. M., Stier, N. and Roth, M. (2000). Preliminary findings from ecological studies of the raccoon dog (*Nyctereutes procyonoides*) in eastern Germany. *Zeitschrift fur Okologie und Naturschutz 9*, 147-152.

Drygala, F., Stier, N., Zoller, H., Bögelsack, K., Mix, H. M. and Roth, M. (2008c). Habitat use of the raccoon dog (*Nyctereutes procyonoides*) in north-eastern Germany. *Zeitschrift für Säugetierkunde 73*, 371-378.

Drygala, F., Stier, N., Zoller, H., Mix, H. M., Bögelsack, K. and Roth, M. (2008a). Spatial organisation and intra-specific relationship of the raccoon dog *Nyctereutes procyonoides* in Central Europe. *Wildlife Biology 14*, 457-466.

Drygala, F., Zoller, H., Stier, N., Mix, H. and Roth, M. (2008b). Ranging and parental care of the raccoon dog *Nyctereutes procyonoides* during pup rearing. *Acta Theriologica 53*, 111-119.

Drygala, F., Zoller, H., Stier, N. and Roth, M. 2010. Dispersal of the raccoon dog *Nyctereutes procyonoides* into a newly invaded area in Central Europe. *Wildlife Biology 16*, 150-161.

Helle, E. and Kauhala, K. (1991). Distribution history and present status of the raccoon dog in Finland. *Holarctic Ecology 14*, 278-286.

Helle, E. and Kauhala, K. (1993). Age structure, mortality, and sex ratio of the raccoon dog in Finland. *Journal of Mammalogy 74*, 936-942.

Helle, E. and Kauhala, K. (1995). Reproduction of the raccoon dog in Finland. *Journal of Mammalogy 76*, 1036-1046.

Holmala, K. and Kauhala, K. (2006). Ecology of wildlife rabies in Europe. *Mammal Review 36*, 17-36.

Holmala, K. and Kauhala, K. (2009). Habitat use of medium-sized carnivores in southeast Finland - Key habitats for rabies spread? *Annales Zoologici Fennici 46,* 233-246.

Ihalainen, A. (2012). [*Diet of raccoon dogs and badgers in Ruissalo Island, Turku.*] – Master thesis, University of Turku, Turku (In Finnish).

Ikeda, H. (1983). Development of young and parental care of the raccoon dog *Nyctereutes procyonoides viverrinus* Temmick, in captivity. *Journal of the Mammalian Society of Japan 9,* 229-236.

Ikeda, H. (1984). Raccoon dog scent marking by scats and its significance in social behaviour. *Journal of Ethology 2*, 77-84.

Ivanova, G. I. (1962). [Comparison of the diets of the red fox, badger and raccoon dog in Voronez national Park.] *Ucenye Zapiski/Moskovskij Gosudarstvennij Pedagogičeskij Institut im. V. I. Lenina 186*, 210-256 (In Russian).

Jędrzejewska, B. and Jędrzejewski, W. (1998). Predation in Vertebrate Communities: the Białowieża Primeval Forest as a case study. *Ecological Studies (Springer Verlag) 135*, 215-219.

Judin, V. G. (1977). [Raccoon dog in Primorje and Priamurye.] *Nauka, Moscow*, Pp. 76-98 (In Russian).

Kauhala, K. (1992). *Ecological characteristics of the raccoon dog in Finland.* Ph.D. thesis, University of Helsinki, Helsinki.

Kauhala, K. (1993). Growth, size, and fat reserves of the raccoon dog in Finland. *Acta Theriologica 38*, 139-150.

Kauhala, K. (1996a). Reproductive strategies of the raccoon dog and the red fox in Finland. *Acta Theriologica 41*, 51-58.

Kauhala, K. (1996b). Habitat use of raccoon dogs, *Nyctereutes procyonoides*, in southern Finland. *Zeitschrift für Säugetierkunde 61,* 269-275.

Kauhala, K. (2004). Removal of medium-sized predators and the breeding success of ducks in Finland. *Folia Zoologica 53*, 367-378.

Kauhala, K. (2007). [The numbers of medium-sized carnivores in Finland.] *Riista- ja kalatalous – Selvityksiä 1*, 1-18 (In Finnish).

Kauhala, K. (2009). [Diet of the omnivorous raccoon dog in Europe and in the Far East.] *Suomen Riista 55*, 45-62 (In Finnish with English summary).

Kauhala, K. and Auniola, M. (2001). Diet of raccoon dogs in summer in the Finnish archipelago. *Ecography 24*, 151-156.

Kauhala, K. and Auttila, M. (2010). Habitat preferences of the native badger and the invasive raccoon dog in southern Finland. *Acta Theriologica* 55: 231-240.

Kauhala K. and Helle, E. (1994). [Home ranges and monogamy of the raccoon dog in southern Finland.] Suomen Riista 40, 32-41 (In Finnish with English summary).

Kauhala, K. and Helle, E. (1995). Population ecology of the raccoon dog in Finland – a synthesis. *Wildlife Biology 1*, 3-9.

Kauhala, K., Helle, P. and Helle, E. (2000). Predator control and the density and reproductive success of grouse populations in Finland. *Ecography 23*, 161-168.

Kauhala, K., Helle, E. and Taskinen, K. (1993b). Home range of the raccoon dog (*Nyctereutes procyonoides*) in southern Finland. *Journal of Zoology (London) 231*, 95-106.

Kauhala, K. and Holmala, K. (2006). Contact rate and risk of rabies spread between medium-sized carnivores in southeast Finland. *Annales Zoologici Fennici 43*, 348-357.

Kauhala, K. and Holmala, K. (2008). Optimal radio-tracking strategy - The best results with the least effort. *Acta Theriologica 53*, 333-341.

Kauhala, K. and Holmala, K. (2011). Landscape features affect home range size of northern badgers (*Meles meles*). *Annales Zoologici Fennici 48*, 221-232.

Kauhala, K., Holmala, K., Lammers, W. and Schregel, J. (2006). Home ranges and densities of medium-sized carnivores in south-east Finland, with special reference to rabies spread. *Acta Theriologica 51*, 1-13.

Kauhala, K., Holmala, K. and Schregel, J. (2007). Seasonal activity patterns and movements of the raccoon dog, a vector of diseases and parasites, in southern Finland. *Mammalian Biology 72*, 342-353.

Kauhala, K., Kaunisto, M. and Helle, E. (1993a). Diet of the raccoon dog, *Nyctereutes procyonoides*, in Finland. *Zeitschrift für Säugetierkunde 58*, 129-136.

Kauhala, K. and Kowalczyk, R. (2011). Invasion of raccoon in dog in Europe: history of introduction, colonization, and threat to the native fauna. *Current Zoology 55*, 584-598.

Kauhala, K., Laukkanen, P. and von Rege, I. (1998b). Summer food composition and food niche overlap of the raccoon dog, red fox and badger in Finland. *Ecogaphy 21*, 457-463.

Kauhala, K., Pietilä, H. and Helle, E. (1998c). Time allocation of male and female raccoon dog to pup rearing at the den. *Acta Theriologica 43*, 301-310.

Kauhala, K. and Saeki, M. (2004). Finnish and Japanese raccoon dogs: On the road to speciation? In: Macdonald, D. W. and Sillero-Zubiro, C. (eds.), *Biology and Conservation of Wild Canids*. Oxford University Press, Oxford, Pp. 217-226.

Kauhala, K. and Salonen, L. (2012). Does a non-invasive method – latrine surveys – reveal habitat preferences of raccoon dogs and badgers? *Mammalian Biology* (in press). Doi: 10.1016/j.mambio.2012.02.007.

Kauhala, K., Schregel, J. and Auttila, M. (2010). Habitat impact on raccoon dog *Nyctereutes procyonoides* home range size in southern Finland. *Acta Theriologica 55,* 371-380.

Kauhala, K., Viranta, S., Kishimoto, M., Helle, E. and Obara, I. (1998a). Skull and tooth morphology of Finnish and Japanese raccoon dogs. *Annales Zoologici Fennici 35*, 1-16.

Kobylińska, J. (1996). The red fox and raccoon dog in wetlands of the Biebrza River Valley – Food composition and burrow use. *European Journal of Wildlife Research 1*, 186-189.

Kolosov, A. M., Lavrov, N. P. and Naumov, S. P. (1965). [Biology of game mammals in the Soviet Union.] *Vysšaâ škola, Moscow*, Pp. 69-76 (In Russian).

Korhonen, H. (1987). Energy metabolism of raccoon dog (*Nyctereutes procyonoides*, Gray 1834): applied perspective to common farming practices. *Publications of the University of Kuopio. Original Reports 1/1987,* 1- 69.

Kowalczyk, R., Bunevich, A. N. and Jędrzejewska, B. (2000). Badger density and distribution of setts in Białowieża Primeval Forest (Poland and Belarus) compared to other Eurasian populations. *Acta Theriologica 45*, 395-408.

Kowalczyk, R., Jędrzejewska, B., Zalewski, A. and Jędrzejewski, W. (2008). Facilitative interactions between the Eurasian badger (*Meles meles*), the red fox (*Vulpes vulpes*) and the invasive raccoon dog (*Nyctereutes procyonoides*) in Białowieża Primeval Forest, Poland. *Canadian Journal of Zoology 86*, 1389-1396.

Kowalczyk, R. and Zalewski, A. (2011). Adaptation to cold and predation - shelter use by invasive raccoon dogs *Nyctereutes procyonoides* in Białowieża Primeval Forest (Poland). *European Journal of Wildlife Research 57*, 133-142.

Kowalczyk, R., Zalewski, A., Jędrzejewska, B., Ansorge, H. and Bunevich, A. N. (2009). Reproduction and mortality of invasive raccoon dogs (*Nyctereutes procyonoides*) in the Białowieża Primeval Forest (eastern Poland). *Annales Zoologici Fennici 46*, 291-301.

Lavrov, P. (1971). [Results of raccoon dog introductions in different parts of the Soviet Union.] *Trudy Kafedry Biologii MGZPI 29*, 101-160. (In Russian).

Machnicka-Rowińska, B., Rocki, B., Dziemian, E. and Kołodziej-Sobocińska, M. (2002). [Raccoon dog (*Nyctereutes procyonoides*) - the new host of *Echinococcus multilocularis* in Poland.] *Wiadomosci parazytologiczne 48*, 65-68 (In Polish).

Melis, C., Herfindal, I., Kauhala, K., Andersen, R. and Hogda, K. A. (2010). Predicting animal performance through climate and plant phenology variables: The case of an omnivore hibernating species in Finland. *Mammalian Biology 75*, 151-159.

Mikkola, M. (2011). [A plan for raccoon dog population management.] Suomen riistakeskus, Pohjois-Häme, 23 pp (in Finnish].

Moks, E., Saarma, U. and Valdman, H. (2005). *Echinococcus multilocularis* in Estonia. *Emerging Infectious Diseases 11*, 1973-1974.

Mustonen, A.-M., Asikainen, J., Kauhala, K., Paakkonen, T. and Nieminen, P. (2007). Seasonal rhythms of body temperature in the free-ranging raccoon dog (*Nyctereutes procyonoides*) with special emphasis on winter sleep. *Chronobiology International 24*, 1095-1107.

Naaber, J. (1971). [Raccoon dog.] *Eesti Lodus 14*, 449-455 (In Estonian).

Naaber, J. (1974). [Red fox and raccoon dog in our nature.] In: Merisalu, G. (ed.), *Jaht ja Ulukid. Estonian hunters' handbook 1969-1972*. Eesti Jahimeeste Selts, Tallinn, Pp. 102-115 (In Estonian).

Naaber, J. (1984). [Mammals in Maatsalu today and in the future.] In: Paakspuu V ed. *Eesti NSV Riiklike LooduskaitsealadeTeaduslikud Tööd IV. Maatsalu loodusest*. Valgust, Tallinn (In Estonian).

Nasimovič, A. A. and Isakov, J. A. (eds.) (1985). [Arctic fox, red fox and raccoon dog: distribution of populations, ecology and preservation.] Nauka, Moscow, Pp. 116-145 (In Russian).

Nieminen, P., Mustonen, A.-M., Asikainen, J. and Hyvärinen, H. (2002). Seasonal weight regulation of the raccoon dog (*Nyctereutes procyonoides*): Interactions between melatonin, leptin, ghrelin, and growth hormone. *Journal of Biological Rhythms 17*, 155-163.

Nowak, E. (1973). [Colonization of the raccoon dog (*Nyctereutes procyonoides* Gray) in Europe.] *Beiträge zur Jagd- und Wildforschung 8*, 351-384 (In German).

Nowak, E. and Pielowski, Z. (1964). [Distribution of the raccoon dog in Poland related to its immigration and spread in Europe.] *Acta Theriologica 9*, 81-110 (In German).

Oivanen, L., Kapel, C. M. O., Pozio, E., La Rosa, G., Mikkonen, T. and Sukura, A. (2002). Associations between *Trichinella species* and host species in Finland. *Journal of Parasitology 88*, 84-88.

Oksanen, A. and Lavikainen, A. (2004). Echinococcosis in Fennoscandia. *International Arch. Hylatidosis 35*, 62.

Opermanis, O., Mednis, A. and Bauga, I. (2001). Duck nests and predators: interaction, specialisation and possible management. *Wildlife Biology 7*, 87-96.

Osterman Lind, E., Juremalm, M., Christensson, D., Widgren, S., Hallgren, G., Ågren, E. O., Uhlhorn, H., Lindberg, A., Cedersmyg, M. and Wahlström, H. (2011). First detection of *Echinococcus multilocularis* in Sweden, February to March 2011. *Euro Surveill. 2011 16(14)*:pii=19836. Available online:http://www.eurosurveillance.org/ViewArticle.aspx?ArticleId=19836.

Pitra, C., Schwarz, S. and Fichel, J. (2010). Going west: Invasion genetics of the alien raccoon dog *Nyctereutes procyonoides* in Europe. *European Journal of Wildlife Research 56*, 117-129.

Romig, T., Dinkel, A. and Mackenstedt, U. (2006). The present situation of echinococcosis in Europe. *Parasitology International 55*, S187-S191.

Schwarz, S., Sutor, A., Staubach, C., Mattis, R., Tackmann, K. and Conraths, F. J. (2011). Estimated prevalence of *Echinococcus multilocularis* in raccoon dogs *Nyctereutes procyonoides* in northern Brandenburg, Germany. *Current Zoology 57,* 655-661.

Shibata, F. and Kawamichi, T. (1999). Decline of raccoon dog populations resulting from sarcoptic mange epizootics. *Mammalia 63,* 281-290.

Sidorovich, V. E., Polozov, A. G., Lauzhiel, G. O. and Krasko, D. A. (2000). Dietary overlap among generalist carnivores in relation to the impact of the introduced raccoon dog *Nyctereutes procyonoides* on native predators in northern Belarus. *Zeitschrift für Säugetierkunde 65*, 271-285.

Siemionek, J., Szweda, W. and Radkowski, M. (2007). Epizootic situation of rabies in Warmińsko-Mazurskie Voivodeship in the years 2001-2005. *Medycyna Wweterynaryjna 63*, 700-702.

Singer, A., Kauhala, K., Holmala, K. and Smith, G. C. (2009). Rabies in Northeastern Europe – the threat from invasive raccoon dogs. *Journal of Wildlife Diseases 45*, 1121-1137.

Sutor, A. (2008). Dispersal of the alien raccoon dog *Nyctereutes procyonoides* in Southern Brandenburg, Germany. *European Journal of Wildlife Research 54*, 321-326.

Sutor, A., Kauhala, K. and Ansorge, H. (2010). Diet of the raccoon dog (*Nyctereutes procyonoides*) – a canid with an opportunistic foraging strategy. *Acta Theriologia 55*, 165-176.

Thiess, A., Schuster, R., Nockler, K. and Mix, H. (2001). [Helminth findings in indigenous raccoon dogs *Nyctereutes procyonoides* (Gray, 1843).] *Berl. Münch. Tierärzlt. Wschr. 114*, 273-276 (In German with English summary).

Viranta, S. and Kauhala, K. (2011). Increased carnivory in Finnish red fox females – adaptation to a new competitor? *Annales Zoologici Fennici 48*, 17-28.

Viro, P. and Mikkola, H. (1981). Food composition of the raccoon dog *Nyctereutes procyonoides* Gray, 1834 in Finland. *Zeitschrift für Säugetierkunde 46*, 20-26.

Westerling, B. (1991). [Rabies in Finland and its control 1988-90.] *Suomen Riista 37*, 93-100 (In Finnish with English summary).

WHO (2002). Summarizing the rabies situation in Europe 1990-2002 from the Rabies Bulletin Europe. *Rabies Bulletin Europe 26(4)*, 11-17.

WHO (2003). Distribution of rabies in Europe. *Rabies Bulletin Europe 27(4)*, 8-22.

WHO (2005). Distribution of rabies in Europe. *Rabies Bulletin Europe 29(4)*, 7-20.

WHO (2009). Distribution of rabies in Europe. *Rabies Bulletin Europe 33(4)*, 9-20.

WHO (2010). Distribution of rabies in Europe. *Rabies Bulletin Europe 34(4)*, 12-24.

Yamamoto, I. (1984). Latrine utilization and feces recognition in the raccoon dog, *Nyctereutes procyonoides*. *Journal of Ethology 2*, 47-54.

Yamamoto, I. (1987). Male parental care in the raccoon dog *Nyctereutes procyonoides* during the early rearing period. In: Itô, Y., Brown, J. L. and Kikkawa, J. (eds.), *Animal Societies: Theories and Facts*. Japan Sci. Soc. Press, Tokyo, Pp. 189-195.

In: Carnivores
Editors: F. I. Álvares and G. E. Mata
ISBN 978-1-62257-323-3

Chapter 3

A Comparative Analysis of Human Perceptions of Cougars and other Large Carnivores in Canada and El Salvador

Michael O'Neal Campbell*
Department of Geography, Thompson Rivers University,
Kamloops, BC, Canada

Abstract

The relationship between people and the cougar *Puma concolor* and other larger predators in Canada (brown bears *Ursus arctos*, black bears *Ursus americanus*) and Latin America (jaguars *Panthera onca*) is examined in a comparative mode for cases in Kamloops and Vancouver Island, Canada and El Salvador. Literature derived hypotheses mooting greater sympathy for carnivores from women and young people, greater fear from women and older people, and greater fear of the larger brown bears and jaguars over cougars are tested. The results show that the hypotheses are both supported and refuted. The size of the predator is as important as common ideas about its behaviour, and troublesome invasive behaviour (mostly from bears) is at least as important as predatory behaviour (largely from cougars and jaguars). However, consistent with hypotheses, women and older people were in many

* Email: ecol55@hotmail.com.

instances more afraid of cougars and other carnivores, and older people were generally less tolerant of carnivores. Differences included opinions on positive impacts on quality of life and useful functions of cougars (favoured by men in El Salvador, both men and women in Vancouver Island, and women in Kamloops), and danger to adults, children and other animals (favoured by women more than men in El Salvador for all groups, women more than men in Vancouver Island for children and animals, not adults, and women more than men in Kamloops for adults and children, not animals). For age, older people in El Salvador were generally less tolerant than younger people of large carnivores, similar to Vancouver Island, but not Kamloops, where there were no age-related opinions of the danger from carnivores towards people. The cougar's elusive behaviour limited comparisons with bears and jaguars. Awareness of such issues may assist wildlife conservation.

INTRODUCTION

Large, wild carnivores hold a special and unique position in the conservation debate, because they contribute to strongly negative attitudes among people, especially in rural areas, due to fear of attacks on humans and/or livestock, resultant restrictions on human life spaces, and the monetary, life and lifestyle losses incurred as a result (Saenz and Carrillo 2002; Conforti et al. 2003; Kleiven et al. 2004; Hoogesteijnal and Hoogesteijnal 2008). The study of large carnivores is also of practical relevance, because wide-ranging nocturnal carnivores are frequently used as umbrella species in conservation planning (Soule and Terborgh 1999; Maehr et al., 2001, 2002; Carroll et al. 2003).

As such their relations with people influence habitat and landuse planning, and conservation policy (Dickson and Beier 2007). Debateable issues concern the actual inter-species based factors for human attitudes towards particular species, and the relevance of these attitudes for conservation possibilities (Campbell and Lancaster 2010).

In all continents, different species of large carnivores share landscapes and spaces of conflict with people, with examples being: brown bears *Ursus arctos* and wolves *Canis lupus* in Eurasia (Schwartz et al. 2003); tigers *Panthera tigris* and leopards *Panthera pardus* in India and South east Asia (Johnsingh 1983; Karanth and Sunquist 1995; Johnsingha and Negib 2003); lions *Panthera leo,* leopards and cheetahs *Acinonyx jubatus* in sub-Saharan Africa (Maclennan et al 2009); cougars *Puma concolor*, black bears *Ursus americanus* and brown bears *Ursus arctos* in North America (Kellert et al. 1996; Decker et al. 2001); and cougars and jaguars in Latin America (Taber 1997; Scognamillo et al. 2003; Campbell 2011).

In such scenarios of conservation, conflict and strategic possibilities, important issues concerning inter-species variables are the aggressiveness, perceived or real, population density, landcover preferences, size, and reputation of the resident carnivores, these being important factors for public attitudes and hence conservation efforts (Treves and Karanth 2003; Hayward and Somers 2009; Inskip and Zimmermann 2009; Campbell and Lancaster 2010). For example, in North America several studies have found that due to interspecies differences in size and behaviour, people generally have more positive opinions about black bear presence than that of the cougar, because of the perception that the latter animal is more predatory (Davis et al. 2001; Beier, 1991).

Brown bears are more feared than black bears, because of their larger size and reputed aggressiveness (Kellert, 1994; Pelton et al. 1976). Furthermore, people are more tolerant of black bears than of cougars and brown bears, and of cougars than brown bears (Pelton et al. 1976; Hayward and Somers, 2009; Deurbrouck, 2007; Thornton and Quinn, 2009; Kellert, 1994)

Concerning human variables, in the comparative analysis of human attitudes to two or more species of predators, the respondents' gender and age are particularly important (Campbell and Lancaster 2010). Several hypotheses regarding gender and age related opinions towards large carnivores have been posed in the literature. As argued by Weber and Rabinowitz (1996), Kleiven et al. (2004) and Campbell and Lancaster (2010), women may be more tolerant of large carnivores than men, but express more fear and more concern for child safety.

These hypotheses have been examined in the North American context (Morzillo et al. 2007; Bowman et al. 2001, 2004; Decker et al. 2001; Kellert et al. 1996). Kellert (1994) and Williamson (2002) hint that demographic characteristics such as age and gender may contribute to different attitudes toward large carnivores, with women and older people being more concerned because they feel more vulnerable.

Women are more afraid of bears than men are of these animals (Morzillo et al. 2010). Women and younger people are more likely than men and their elders, to support carnivore conservation by choosing toleration or trapping and removal rather than shooting the animals (Fulton et al. 1995; Kellert, 1976; 1989; Morzillo et al. 2007; Teel et al. 2002; Hayward and Somers, 2009; Deurbrouck, 2007; Thornton and Quinn, 2009)

The Case of the Cougar *Puma concolor* and the Larger Sympatric Carnivores

Few carnivores have as wide a range as the cougar, also known as the puma, mountain lion and panther, this species being found through most of North and South America (Campbell and Lancaster 2010). Throughout its range, the cougar shares its habitats with people, and also with larger carnivores (brown bears *Ursus arctos* and black bears *Ursus americ*anus in North America, and jaguars *Panthera onca* in Mexico southwards (Taber 1997; Nunez et al. 2000, 2002; Decker et al. 2001; Crawshaw and Quigley, 2002; Oliveira, 2002; Campbell and Lancaster 2010, Campbell and Torres Alvarado 2011). The cougar is generally perceived as an elusive, secretive carnivore, with few records of attacks on people, moderate danger to livestock, but fear of attacks on people, mostly children (Beier, 1991, 1993; Turner et al. 1990; Herbert, 1989). Cougars, similar to the larger carnivores, are frequently situated in scenarios of conflict with people, due to human population expansion, landscape change and habitat conversion, increased livestock presence and perceived danger (Campbell and Torres Alvarado 2011). These conflicts are especially important for carnivore survival, because human comparisons of different species strongly affect the conservation possibilities of different species (Treves and Karanth 2003; Hayward and Somers 2009; Inskip and Zimmermann 2009).

There are both positive and negative attitudes towards cougars, the positive views based mainly on the cougar's status as a wildlife symbol and its relation to the domestic cat (Deurbrouck, 2007; Thornton and Quinn, 2009), and the negativity based on their predation threat to pets and children (McKee, 2003). Few studies have compared the public attitudes towards cougars in widely distributed parts of its range, and included a comparison with other predators in these varied regions, although some studies have compared public attitudes and ecological relations of cougars and bears in North America (Kellert et al. 1996; Decker et al. 2001; Teel et al. 2002) and similar topics on cougars and jaguars in Central and South America (Jorgenson and Redford 1993; Taber et al. 1997; Farrell 1999; Polisar 2000, 2003; Conforti et al. 2003; Scognamillo et al. 2003; Monroy-Vilchis 2009).

Key questions and hypotheses that would emerge from such a comparative study (necessarily based on several literature sources, as the vastness of the cougar's range would make a field study difficult) would concern the importance of the animals' physical size (in this example, the cougar is generally smaller than the bears and the jaguar), and behaviour, as the cougar is generally seen as more elusive (even though the jaguar and brown bear may generally be reported as rarer), less aggressive than the other animals, less of a threat to livestock and humans, except to children, and less invasive than bears. Also important would be the human variables, generally gender and age, as these are all pervasive, and in some cases education and employment, even though these latter variables may be less important in the rural contexts where agriculture is the main livelihood activity. This paper analyses the results of three articles that attempt a comparison of cougars, bears and jaguars (Campbell and Lancaster 2010, Campbell and Torres Alvarado 2011 and Campbell 2012) based on Vancouver Island and Kamloops in Canada and El Salvador in Central America, and evaluates their findings within the broader context of the literature on the specific species, the sociology of human gender and age and general human relations with large carnivores.

The Cougar and the Jaguar

A major factor for the different and similar attitudes to jaguars and cougars is their size, and consequently the size and type of prey taken by these cats (Campbell and Torres Alvarado 2011). Jaguars are described as the third largest cat species (after tigers and lions), with some male jaguars weighing as much as female tigers or lions (up to 160 kg) especially in the Amazon forests, but averaging 56–96 kg (Burnien and Wilson 2001). Cougars are described as the fourth largest cats, marginally larger than leopards, but varying greatly in size (for example cougars in the colder regions of Canada are described as larger than some jaguars (75 kg) (Polisar 2000, 2003). Their size range is about 29 - 100 kg (males typically weigh 53 to 100 kilograms) (Nowell and Jackson 2006). Strong positive correlations have been described between mean prey mass and cougar (Iriarte et al. 1990) and jaguar size (Oliveira 1994). According an analysis by Oliveira (1994) the jaguar:cougar body ratio was more equal in Central (1:0.9) than South America (1:0.6), similar to the mean prey weight ratio in Central (1:0.9) and South America (1:0.6), hinting that cat size affects the size of prey taken. Studies have found that jaguars kill larger mammals than cougars, in some cases the prey of the jaguar weighing twice as much as that of the cougar (Oliveira. 2002; Palmeira et al. 2008).

These characteristics provide evidence for the relations between these cats and human life spaces. In terms of domestic animals, cattle have been described as important targets, but jaguars may focus on adult animals while cougars take calves, and smaller animals such as sheep and goats (Yàñez et al. 1986; Boulhosa and Valdes, 1997; Oliveira. 2002). Neither the jaguar nor the cougar commonly kills people (unlike tigers, lions and leopards in other continents), despite fears to the contrary in North America for the cougar and mostly South America for the jaguar (Conforti et al. 2003. Campbell and Lancaster 2010, Campbell and Torres Alvarado 2011). In areas of joint distribution of cougars and jaguars, the latter cat is more feared, more blamed for livestock and human attacks and more known, even when as noted by Conforti et al. (2003) these judgements are not justified by evidence (Perry 1970; Guggisberg 1975; Almeida 1990; Beier 1991; Foerster 1996). The cougar is less known than the jaguar, due to its elusive nature (Conforti et al. 2003; Campbell 2010).

The results of my study (Campbell and Torres Alvarado 2011) of jaguars and cougars in El Salvador are particularly interesting, as these animals are largely extinct in that country (but appear to cross borders), and hence reintroductions are an important topic. The jaguar and cougar remain in the public consciousness due to livestock kills and threats to people (Conforti and Azevedo 2003). The relatively small size of El Salvador facilitates animal movements and prevents final judgements on extinction. El Salvador (population 6,052,064, area 21 041 square kilometres, population density 288 per square kilometre; Central Intelligence Agency 2010, Figure 1) is the most densely populated country in Central and South America (Hecht et al. 2006). In this study four hypotheses were posed, based on the literature on human attitudes to these animals (Weber and Rabinowitz 1996; Kleiven et al. 2004; Campbell and Lancaster 2010). These were: (1) women are more tolerant of large carnivores than men; (2) women are more afraid of large carnivores; (3) both men and women are more afraid of jaguars than cougars; and (4) older people are less tolerant of large carnivores than younger people. For a comparison of the cougar and the jaguar, the most relevant was hypothesis 3.

The results showed that there were no significant differences in attitudes towards the two species in terms of toleration, trapping and removal, shooting, protection, reintroductions, impacts on human life, or danger to cattle, human adults or children. As explained by some respondents, any difference in strength between the jaguar and cougar would make little difference to a man, woman or child in an encounter. However, more people thought cougars were more dangerous than jaguars (especially women, although this did not affect the other assessments). This runs against the common perception of the jaguar as a powerful cattle killer, and the cougar as the more elusive, timid animal, in some cases not even dangerous to people (Conforti and Azevedo 2003;

Palmeira et al. 2008). It nevertheless supports the findings of the study by Conforti and Azevedo (2003) on the comparative local fears and perceptions of jaguars and cougars, which found that although fewer people feared the cougar than the jaguar, feelings towards the two species were similar: people with positive perceptions towards jaguars felt the same towards cougars, and people with a negative opinion of jaguars tended to have the same opinion of cougars, regardless of actual behavioural differences between the two cats.

Figure 1. Location of El Salvador.

The Cougar and the Brown Bear

The brown bear is the largest mammalian carnivore in the Americas, weighing from 100 to 322 kg, with most of the larger bears found in coastal regions of British Columbia and Alaska (McLellan 1994).

In the literature, brown bears are more feared, with a more 'nefarious image,' than cougars (Kellert et al.1996)., and are also more feared than black bears, because of their superior size and fierce reputation (Kellert 1994; Pelton et al. 1976), despite the generally positive opinion that North Americans have of bears (Hayward and Somers 2009; Kellert 1994), and the perception of the cougar as a predation threat to pets and children (McKee, 2003). In my study on attitudes to brown bears, black bears and cougars in Kamloops, British Columbia (Campbell, forthcoming for 2012 in the *Journal of Society and Animals,* Figure 2), based on the literature that found that gender and age related differences in support for bear and cougar conservation (Teel et al. 2002) and fear of bears and cougars (Kellert, 1989, 1994; Thornton and Quinn, 2009; Morzillo et al. 2010) the following hypotheses were posed: (1) women and older people are more afraid of bears and cougars than men and younger people, respectively; (2) women and younger people are more likely than men and their elders, respectively, to support bear and cougar conservation by choosing toleration or trapping and removal rather than shooting the animals; and (3) more people will be tolerant of black bears than of cougars and brown bears, and of cougars than brown bears. The evidence for the third hypothesis, showed that cougars were seen as the most dangerous to children and pet animals, followed by brown bears. Brown bears were seen as the most dangerous to adults, followed by cougars.

The Cougar and the Black Bear

The black bear is generally larger than the cougar (41 - 250 kg) (Brown 1993). In areas of joint distribution, cougars are perceived as more dangerous than black bears, the main indicators being that more people (especially men) approve of cougar hunting than black bear hunting (Teel et al. 2002). In my paper on black bears, cougars and brown bears in Kamloops (Campbell 2012), black bears were seen as the least dangerous of the three animal species. In my study comparing cougars and black bears in Vancouver Island, Canada (Campbell and Lancaster 2010, Figure 3), based on the findings of the literature on public perceptions of black bears and cougars in North America (Morzillo et al., 2007; Bowman et al.,2001, 2004; Decker et al., 2001; Kellert et al., 1996), the following hypotheses were investigated: (a) people generally have positive more opinions about black bear than cougar presence, because of the perception that the latter animal is more predatory (Davis et al. 2001; Beier 1991); (b) demographic characteristics such as age and gender may contribute to different attitudes toward large carnivores, with women and older people being more concerned because they feel more vulnerable (Williamson 2002; Kellert 1994).

In terms of the first hypothesis, few responses indicated any species-specific differences in people's perceptions, suggesting that people view the two predators similarly. Cougars were perceived as more dangerous overall than bears, especially toward children and companion/farm animals, but bears were perceived as engaging in more troublesome behaviours than cougars. Attitudes about how to treat the two predators were essentially identical, but there were differences on the reasons. Reasons for shooting cougars comprised the perception of their predatory nature (compared with the more omnivorous bears) and their greater likelihood to attack pets (dogs, cats, and birds), livestock, and children.

Figure 2. Kamloops, in Canada.

Reasons for shooting bears mostly concerned. their invasion of private property or social spaces. Similar to these findings, Morzillo et al. (2007) point out that, at a general level "people tend to respond quite similarly to different large carnivore species regardless of ecological and behavioural differences" (p. 418). This point is supported by several studies (Kleiven et al., 2004; Kellert et al., 1996; Kellert, 1985).

THE IMPORTANCE OF GENDER

According to literature sources, North American women are more afraid than men of cougars (Kellert, 1989, 1994; Thornton and Quinn, 2009) yet women are more supportive than men of cougar conservation (Kellert, 1989; Teel et al. 2002; Thornton and Quinn, 2009). Based on the results in my paper on El Salvador (Campbell and Torres Alvarado 2011), there were important findings based on gender. More women than men were afraid of the big cats. More men than women supported toleration of jaguars and cougars, with no gender differences on trapping and shooting, similar proportions of men and women thought the cats needed more protection, or could be reintroduced into protected areas, general rural areas or zoos.

Slightly more women than men saw a danger to pets, livestock and people, but this was not statistically relevant. More men than women thought jaguars were more dangerous, while due to undecided answers more men than women also thought that cougars were more dangerous than jaguars, but by a smaller margin. More women than men thought the presence of the carnivores made human life worse, and more men than women thought they made life better. Relating these results to the hypotheses, the first hypothesis of the study was that women are more tolerant of large carnivores than men. This hypothesis is supported by eco-feminist, 'women and environment' and the political ecology literature that women, through cultural conditioning and/or natural cognition have a greater sympathy for the environment/wildlife (Campbell 1998, 2005a, 2005b).

This hypothesis was generally rejected, as men were more tolerant of large carnivores, and gender differences in the shooting of the animals were insignificant. Men were also more likely to agree that large carnivores improved human life, while more women thought otherwise. For the second hypothesis, that women are more afraid of wildlife and are more concerned about the danger to children, the results supported the first part but not the second (in contrast to papers on Canada described below).

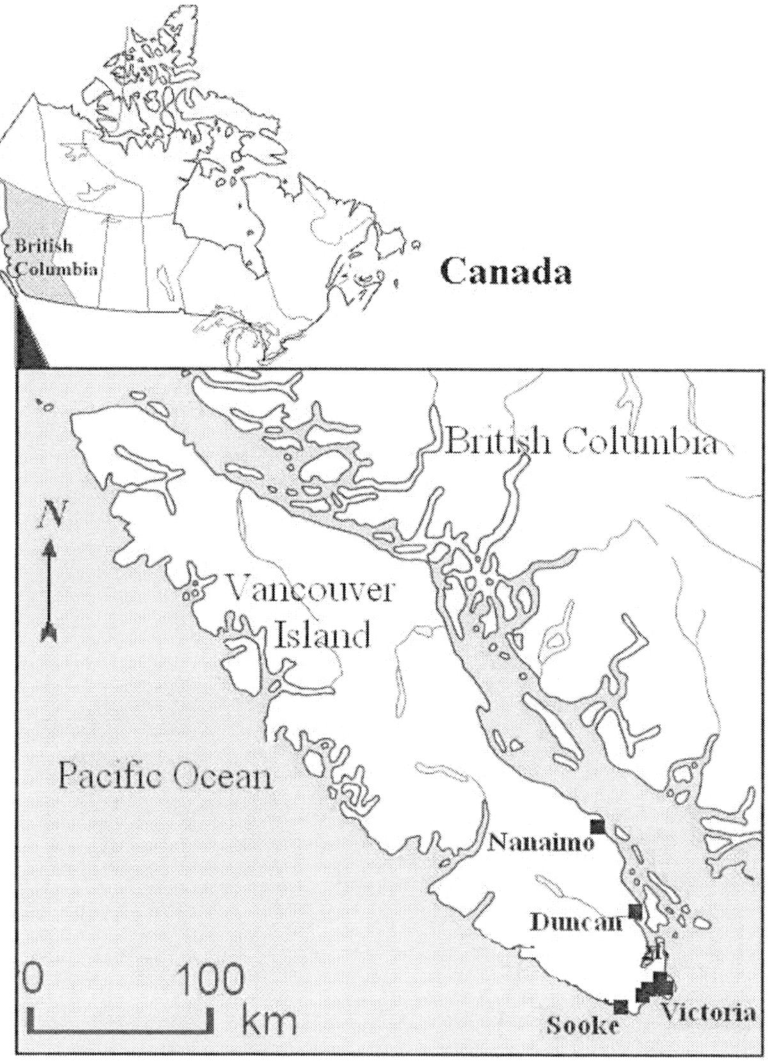

Figure 3. Vancouver Island, in Canada.

Based on the hypotheses of my two Canadian papers, there were also important findings based on gender. In the study on Vancouver Island (Campbell and Lancaster 2010), while men and women differed little on whether to shoot, trap or remove cougars and black bears, more women than men thought that carnivores were dangerous to pet animals and livestock and children. Men and women did not differ on the danger to adults, although more people saw cougars as dangerous than bears. While people thought bears were more likely to engage in troublesome behaviour than cougars, more men than women thought that bears were troublesome (there was no gender distinction for cougars). Most people did not think that carnivores diminished the quality of life, and there were no gender distinctions. In my second study for Kamloops (Campbell 2012), significantly more women than men viewed the carnivores as dangerous to adults and children, partly because many women believed that women were closer to children. In this study, more women than men thought the bears were troublesome, but there was little difference between women and men in their opinion of cougars. There were no differences for gender regarding shooting, trapping or removal of the carnivores. More women than men thought cougars had useful functions, and enhanced human quality of life, but their opinions about bears were similar.

THE IMPORTANCE OF AGE

According to the literature, younger people are more supportive of cougar conservation and less afraid of these animals than their elders (Fulton et al. 1995; Kellert, 1976; 1989; Teel et al. 2002). Concerning the hypothesis about age in the study of El Salvador (Campbell and Torres Alvarado 2011), the fourth hypothesis was that older people would be less tolerant of wild animals than younger people. However, age was irrelevant to the fear of the animals, toleration, reintroductions to unrestricted rural areas and zoos, but there was greater support among older people for shooting and less support for reintroductions to special areas and a stronger perception of danger to cattle. These findings are partly consistent and inconsistent with other literature on the topic that young people are more tolerant of large carnivores (Andersone and Ozolins 2004) and of environmental issues in general (Batterbury et al. 1997; Campbell 2005b). For the study of Vancouver Island (Campbell and Lancaster 2010), as noted above, the second hypothesis was that human age could contribute to different attitudes toward large carnivores, with women and older people being more afraid because they feel more vulnerable (Williamson 2002, Kellert 1994). The results showed that more younger people and fewer older people supported tolerance of large carnivores, and fewer younger people and more older people supported the trapping and removal of the animals. Although more older people saw bears as a dangerous to pets and livestock, and the reverse was true for danger to adults, there were no significant differences between young and old for cougars, or for the danger of any animal to children. Neither older or younger people believed that cougars were troublesome, while more younger people thought that bears were more troublesome than cougars, and as a consequence were less likely to reduce the quality of life.

For the study of Kamloops (Campbell 2012) the first two hypotheses were relevant to age (older people will be more afraid of bears and cougars, and younger people are more likely to support bear and cougar conservation by choosing toleration or trapping and removal rather than shooting the animals). The results showed that there were no age-related opinions of the danger from carnivores towards people, but more older people saw brown bears as dangerous towards pets. Also, there were no age-related opinions on the troublesome behaviour of cougars, but older people were more likely than younger people to see bears as troublesome. Inconsistent with hypothesis 2, and although most people selected trapping and removal over shooting, more younger people believed cougars and bears should be shot, and more older people believed in trapping and removal. Age-related opinions on the usefulness, and contribution to quality of life, of cougars and bears were insignificant.

COMPARISONS AND CONCLUSION

The results of these three papers, situated within the broader literature review illustrate some important points. Firstly, the cougar was seen as a dangerous animal, only slightly different from the larger carnivores. While broadly similar to the jaguar and brown bear in dangerousness, it was definitely more dangerous, at least in some situations than the black bears. Its potential as a predator appeared more important than its elusive behaviour, the latter characteristic only matched by the troublesome behaviour of bears. Secondly, gender was a very important variable affecting viewpoints, but these were inconsistent across the study areas. In all three cases, more women than men thought the carnivores were dangerous, especially to people but also to other animals, but in terms of the lesser problem of troublesome behaviour and positive contributions to human life, there were contrasting results, especially between the more negative opinions of women in El Salvador and the more positive opinions of those in Kamloops. Regarding age related opinions, which were generally weaker than those related to gender, Salvadorans showed a stronger generational gap, possibly because the near extinction of the big cats elicited more sympathy and regrets from the younger generation. In Canada the generational difference was stronger for bears, possibly due to their greater visibility and more variable behaviour.

Relating the results of this literature review to the points listed as the key questions and hypotheses (the importance of the cougar's physical size and behaviour relative to the those of the jaguar and the bears), the evidence points to lack of specific knowledge of the differences between the cougar and the jaguar, and greater knowledge of bear behaviour, as key issues, these being superimposed on fairly strong but variable gender based opinions and weaker age related opinions. The contradictions between the Canadian examples also show that individual life experiences may be at least as important as generalised attitudes, as the former may skew results. The conclusion, based on these wide ranging but limited examples must also hint that the cougar's reputation as a dangerous animal is possibly larger than reality, easily matching that of the jaguar and the brown bear, and subordinate only to the troublesome behaviour of bears. Further studies may focus on actual relative impacts of these carnivores (measuring attacks on people and other animals, threatening behaviour and impacts of people's movements) for a comprehensive evaluation of the accuracy of social perceptions.

REFERENCES

Almeida, T. 1990. *Hunting in the Mato Grosso and Bolivia*. Safari Press, Long Beach.

Andersone, Z. and Ozolins, J. 2004. Public perception of large carnivores in Latvia. *Ursus*. 15: 181–7.

Batterbury, S, Forsyth, T. and Thomson, K. 1997. Environmental transformations in developing countries: hybrid research and democratic policy. *The Geographical Journal* 163: 126–32.

Beier, P. 1991. Cougar attacks on humans in the United States and Canada. *Wildlife Society Bulletin*. 19: 403- 412.

Beier, P. 1993. Determining minimum habitat areas and habitat corridors for cougars. *Conservation Biology*. 7(1): 94-108.

Boulhosa, R. L. P., and Valdes. S. A. C. 1997. *Impact of big cats predation on domestic cattle in the Pantanal Matogrossense, Brazil*. 4th Research report: September/96-August/97. CENAP/Associação Pró-Carnívoros, São Paulo.

Bowman, J. L., Leopold, B. D., Vilella, F. J., Gill, D. A., and Jacobson, H. A. 2001. Attitudes of landowners toward American black bear compared between areas of high and low bear populations. *Ursus*. 12: 153-160.

Bowman, J. L., Leopold, B. D., Vilella, F. J., and Gill, D. A. 2004. A spatially explicit model,derived from demographic variables, to predict attitudes toward black bear restoration. *Journal of Wildlife Management*. 68(2): 223-232.

Brown, Gary 1993. *The Great Bear Almanac*. Lyons and Burford, New York.

Burnien, D. and Wilson, D. E. 2001. *Animal: The Definitive Visual Guide to the World's Wildlife*. Dorling Kindersley, New York City.

Campbell, M. 1998. Interactions *between biogeography and rural livelihoods in the coastal savanna of Ghana*. Unpublished PhD thesis Imperial College University of London.

Campbell, M. 2005a. Sacred groves for forest conservation in Ghana's coastal savannas? Assessing ecological and social dimensions. *Singapore Journal of Tropical Geography* 26:151–69.

Campbell, M. 2005b. The ecological and social context of mammal hunting in the coastal savanna of Ghana *Geoforum* 36: 667– 80.

Campbell, M. and Lancaster, B. 2010 Public attitudes towards black bears *Ursus americanus* and cougars *Puma concolor* on Vancouver Island. *Journal of Society and Animals* 181: 40–57.

Campbell, M. and Torres-Alvarado, M. 2011. Public perceptions of jaguars *Panthera onca*, pumas *Puma concolor* and coyotes *Canis latrans* in El Salvador. *Area* 43(3): 250–256.

Campbell, M. Forthcoming for 2012, *Journal of Society and Animals*. The relevance of age and gender for public attitudes to brown bears (*Ursus arctos*), black bears (*Ursus americanus*) and cougars (*Puma concolor*) in Kamloops, British Columbia.

Carroll, C., Noss, R.F. and Paquet, P. 2003. Carnivores as focal species for conservation planning in the Rocky Mountain Region. *Ecological Applications* 11: 961–980.

Central Intelligence Agency 2010. The World Factbook: Central America and Caribbean El Salvador (*https://www.cia.gov/library/publications/the-world-factbook/geos/es.html*). Accessed 8 August 2010.

Conforti, V.L. and Azevedo, F.C.C. 2003. Local perceptions of jaguars *Panthera onca* and pumas *Puma concolor* in the Iguacu National Park area south Brazil. *Biological Conservation*. 111: 215–21.

Crawshaw, P. G. and Quigley, H. B. 2002. Jaguar and puma feeding habits. In *El jaguar en el nuevos milenio*. Una evaluacion de su estado, deteccion de prioridades y recomendaciones para la conservaci ′on de los jaguars en America: 223–235. Medellýn, R. A., Chetkiewicz, C., Rabinowitz, A., Redford, K. H., Robinson, J. G., Sanderson, E. and Taber, A. (Eds). Mexico: D. F. Universidad Autonoma de Mexico/Wildlife Conservation Society.

Davis, H., Wellwood, D., and Ciarniello, L. M. 2001. *"Bear smart" community program: Background report*. British Columbia Ministry of Environment, Lands and Parks, Victoria.

Decker D. J., Brown, T. L. and Seimer, W. F. 2001. *Human dimensions of wildlife management in North America*. Wildlife Society, Bethesda.

Deurbrouck, J. 2007. *Stalked by a Mountain Lion: Fear, Fact and the Uncertain Future of Cougars in America*. Falcon Press, Guilford.

Dickson, B. G. and Beier, P. 2007. Quantifying the influence of topographic position on cougar (*Puma concolor*) movement in southern California, USA. *Journal of Zoology*. 271: 270–277.

Foerster, C. 1996. Researcher attacked by puma in Corcovado National Park, Costa Rica. *Vida Silverstre Neotropical* 5: 57 -58.

Fulton, D. C., Pate, J., and Manfredo, M. J. 1995. *Colorado residents attitudes towards trapping in Colorado*. Colorado State University, Fort Collins.

Guggisberg, C.A.W. 1975. *Wild Cats of the World*. Taplinger Publishing, New York.

Hayward M.W. and Somers M.J. 2009. *Reintroduction of top-order predators*. Wiley-Blackwell, Oxford.

Hecht, S.B., Kandel, S., Gomes, I., Cuellar, N. and Rosa, H. 2006. Globalization, forest resurgence, and environmental politics in El Salvador. *World Development* 34: 308–23.

Herbert, D. 1989. *The status and management of cougar in British Columbia*. Proceedings of the Mountain Lion Workshop, 3, 11-14.

Hoogesteijnal, R. and Hoogesteijnal, A. 2008. Conflicts between cattle ranching and large predators in Venezuela: could use of water buffalo facilitate felid conservation? *Oryx*. 42 (1):132-138.

Inskip, C. and Zimmermann, A. 2009. Human–felid conflict: a review of patterns and priorities worldwide. *Oryx* 43 18–34.

Johnsingha, A.J.T. and Negib, A. S. 2003. Status of tiger and leopard in Rajaji–Corbett Conservation Unit, northern India. *Biological Conservation* 111: 385–393.

Kellert, S. R. 1976. Perceptions of animals in American society. *Transactions of the North American Wildlife and Natural Resources Conference*, 41: 533-545.

Kellert, S. R. 1989. Human-animal interactions: A review of American attitudes to wild and domestic animals in the 20th century. In A. Rowan (Ed). *Animals and People Sharing the World* (pp. 137-176). Hanover, NH: University of New England Press.

Kellert, S. R. 1994. Public attitudes towards bears and their conservation. *International conference on Bear Research and Management.* 9: 43-50.

Kellert, S. R., Black, M., Rush, C. R., and Bath, A. J. 1996. Human culture and large carnivore conservation in North America. *Conservation Biology.* 10: 977-990.

Iriarte, J. A., W. L. Franklin, W. E. Johnson, and K. H. Redford. 1990. Biogeographic variation of food habits and body size of the America puma. *Oecologia* 85: 185-190.

Johnsingh, A.J.T., 1983. Large mammalian prey-predators in Bandipur. *Journal of Bombay Natural History Society* 80: 1–57.

Jorgensen, J.P. and Redford, K.H. 1993. Humans and big cats as predators in the Neotropics. pp 367-390 in M.G.N Dunstone and M.L. Gorman, editors. *Mammals as predators. Series Symposia of the Zoological Society.* Oxford University Press, Oxford.

Karanth, K. U. and Sunquist, M. E. 1995. Prey selection by tiger, leopard, and dhole in tropical forests. *Journal of Animal Ecology* 64: 439–450.

Kleiven, J., Bjerke, T., and Kaltenborn, B. P. 2004. Factors influencing the social acceptability of large carnivore behaviours. *Biodiversity and Conservation.* 13: 1647-1658.

Maclennan, S. D., Groom, R. J., Macdonald, D. W. and Frank, L. G. 2009. Evaluation of a compensation scheme to bring about pastoralist tolerance of lions. *Biological Conservation* 142: 2419.

Maehr, D. S., Noss, R. F. and Larkin, J. L. 2001. *Large mammal restoration.* Island Press, Washington DC.

Maehr, D.S., Land, E.D., Shindle, D.B., Bass, O.L. and Hoctor, T.S. 2002. Florida panther dispersal and conservation. *Biological Conservation.* 106, 187–197.

McKee, D. 2003. Cougar attacks on humans: A case report. *Wilderness and Environmental Medicine.* 14:169–173.

Monroy-Vilchis, O., Rodríguez-Soto, C., Zarco-González, M. and Urios, V. 2009. Cougar and jaguar habitat use and activity patterns in central Mexico *Animal Biology* 59: 145–157.

Morzillo, A. T., Mertig, A. G., Garner, N., and Liu, J. 2007. Resident attitudes toward black bears and population recovery in East Texas. *Human Dimensions of Wildlife.* 12: 417–428.

Morzillo, A. T., Mertig, A. G., Hollister, J. W., Garner, N., and Liu, J. 2010. Socioeconomic factors affecting local support for black bear recovery strategies. *Environmental Management*. 45: 1299–1311.

Nowell, K. and Jackson, P. 2006. *Wild Cats. Status Survey and Conservation Action Plan*. IUCN/SSC Cat Specialist Group. IUCN, Gland.

Nuñez, R., Miller, B. and Lindzey, F. 2002). Food habits of jaguars and pumas in Jalisco, Mexico. *Journal of Zoology(London)*. 252: 373–379.

Nuñez, R., Miller, B. and Lindzey, F. 2002. Ecology of jaguars in the Chamela–Cuixmala Biosphere reserve, Jalisco, Mexico. In *El jaguar en el nuevo milenio*. Una evaluacion de su estado, deteccion de prioridades y recomendaciones para la conservacion de los jaguars en America: 107–126. Medellin, R. A., Chetkiewicz, C., Rabinowitz, A., Redford, K. H., Robinson, J. G., Sanderson, E. and Taber, A. (Eds). Mexico, D. F.: Universidad Nacional Autònoma de Mexico/Wildlife Conservation Society.

Oliveira, T. G. de. 1994. *Neotropical cats: ecology and conservation*. EDUFMA, São Luís,

Oliveira, T. G. de 2002. Comparative feeding ecology of Jaguar and Puma in the Neotropics. Ecología comparativa de la alimentación del jaguar y del puma en el neotrópico. Pp. 265-288 in *El jaguar en el nuevo milenio*. R. A. Medellín, C. Equihua, C. L. Chetkiewicz, P. G. Crawshaw Jr., A. Rabinowitz, K. H. Redford, J. G. Robinson, E. W. Sanderson, A. Taber, eds. Fondo de Cultura Económica/Universidad Nacional Autónoma de México/Wildlife Conservation Society, México, D.F. 2002.

Palmeira F.B.L, Crawshaw, P.G., Haddad, C.M., Ferrazd, K.M.P.M.B and Verdadee, L.M. 2008. Cattle depredation by puma *Puma concolor* and jaguar *Panthera onca* in central-western Brazil. *Biological Conservation*. 141: 118–25.

Pelton, M., Scott, C.. and Burghardt, G. M. 1976. Attitudes and opinions of persons experiencing property damage and/or injury by black bears in the Great Smoky Mountains National Park. *International Conference of Bear Research and Management*. 3: 157-167.

Perry, R. 1970. *The World of the Jaguar*. Taplinger Publishing . New York.

Polisar, J. D. 2000. *Jaguars, Pumas, their Prey Base, and Cattle Ranching: Ecological Perspectives of a Management Issue*. PhD dissertation, University of Florida.

Polisara, J. Maxit, I, Scognamillo, D., Farrell, L., Sunquist, M.E., Eisenberg, J.F. 2003. Jaguars, pumas, their prey base, and cattle ranching: ecological interpretations of a management problem. *Biological Conservation.* 109 (2): 297–310.

Sáenz, J. C. and Carrillo, E., 2002. Jaguares depredadores de ganado en Costa Rica: ¿un 982 problema sin solución? In *El Jaguar en El Nuevo Milenio.* ed. R. Medellin, C. Equihua, 983 C. Chetkiewicz, P. Crawshaw, A. Rabinowitz, K. F. Redford, J. Robinson, E. Sanderson 984 and A. Taber, pp. 127-138. Fondo de cultura económica, Universidad Nacional 985 Autónoma de México, Wildlife Conservation Society, México.

Schwartz, C.C., Swenson, J.E. and Miller, S.D. 2003. Large carnivores, moose and humans: a changing paradigm for the 21st century. *Alces* 39: 41-63.

Scognamillo, D., Maxit, I.E., Sunquist, M. and Polisar, J. 2003. Coexistence of jaguar (*Panthera onca*) and puma (*Puma concolor*) in a mosaic landscape in the Venezuelan Llanos. *Journal of Zoology (London).* 259:269–279.

Soule′, M.E. and Terborgh, J. 1999. *Continental conservation.* Island Press, Covelo.

Taber, A.B., Novaro, A.J., Neris, N., and Colman, F.H., 1997. The food habits of sympatric jaguar and puma in the Paraguayan Chaco. *Biotropica* 29: 204–213.

Teel, T. L., Krannich, R. S., and Schmidt, R. H. 2002. Utah stakeholders' attitudes toward selected cougar and black bear management practices. *Wildlife Society Bulletin.* 30: 2-15.

Thornton, C. T., and Quinn, M. S. 2009. Coexisting with cougars: public perceptions, attitudes, and awareness of cougars on the urban-rural fringe of Calgary, Alberta, Canada. *Human–Wildlife Conflicts.* 3(2): 282–295.

Treves A and Karanth K U 2003. Human–carnivore conflict and perspectives on carnivore management worldwide. *Conservation Biology.* 176: 1491–1499.

Turner, J. W., Wolfe, M. L., and Kirkpatrick, J. F. 1990. Seasonal mountain lion predation on a feral horse population. *Canadian Journal of Zoology.* 70: 929-934.

Weber, W. and Rabinowitz, A.R. 1996. A global perspective on large carnivore conservation. *Conservation Biology.* 104: 1046–1054.

Williamson, D. F. 2002. *Status, management, and trade of the American black bear (Ursus americanus) in North America.* World Wildlife Fund, Washington DC.

Yàñez, J. L., Cardenas, J. C., Gezelle, P. and Jaksic, F. M. 1986. Food habits of the southernmost mountain lions (*Felis concolor*) in South America: natural versus livestocked ranges. *Journal of Mammalogy* 67: 604-606.

In: Carnivores
Editors: F. I. Álvares and G. E. Mata
ISBN 978-1-62257-323-3

Chapter 4

CARNIVORES, CONFLICT, AND CONSERVATION: DEFINING THE LANDSCAPE OF CONFLICT

Todd C. Atwood and Stewart W. Breck
USDA-National Wildlife Research Center,
Fort Collins, CO, US

ABSTRACT

Mitigating conflict between humans and large carnivores is one of the most pressing and intractable concerns in conservation. Yet, there has been surprisingly little effort devoted to incorporating risk assessments of conflict in carnivore conservation and land-use planning. Because human-carnivore conflict can have far-reaching societal and environmental impacts, attention to the 'conflict–conservation nexus' should become integrated into national and global environmental policy-making. However, *how* 'the nexus' is defined, elucidated, and ultimately utilized to forecast and mitigate conflict remains under-explored. Here, we discuss the limitations of current knowledge and methodologies available to forecast human–carnivore conflict and suggest a novel heuristic framework that integrates ecological and sociological data to better predict and mitigate conflict, and optimize conservation planning. We illustrate the utility of our approach using a case study of carnivore connectivity planning in the southwestern United States. Our approach holds promise as an effective tool for use in carnivore conservation by allowing decision-makers to prioritize planning efforts by integrating biological suitability, threat of conflict, and societal acceptance.

INTRODUCTION

Carnivores, particularly top predators, fill vital roles in ecosystems such as contributing to the maintenance of biodiversity (Dalerum et al. 2008), limiting the number of prey species, and functioning as conservation surrogates for less charismatic sensitive species (e.g., Dalerum et al. 2008). Throughout the world, maintaining many populations of large carnivores will require that animals exist in multi-use landscapes in which people are a component of, or the dominant feature on, the landscape. However, where humans and carnivores coexist, competition for shared resources such as prey species or livestock often results in conflict (Thirgood et al. 2000, Sillero-Zubiri and Laurenson, 2001), which we define as a perceived negative interaction between humans and wildlife that results in the implementation of management to reduce the negative interactions. Conflict can have meaningful negative impacts to people and the management of conflict animals can be detrimental to conservation efforts. Indeed, anthropogenic factors including conflict with humans are the primary driver of global declines in several large carnivore species such as African lions (*Panthera leo*), tigers (*Panthera tigris*), and Mexican wolves (*Canis lupus baileyii*) (Michalski et al. 2006). Faced with these issues, resolving conflicts between people and predators is of fundamental importance to developing effective conservation strategies for large carnivores.

Human-wildlife conflict is distinct from typical biological parameters (e.g., animal behavior, population dynamics, or species richness) in that it is as much a sociological phenomenon as it is a biological phenomenon. Thus people with differing beliefs and attitudes towards wildlife and the actions of wildlife can influence the perception of what is or is not deemed conflict. For example, some cultures have greater tolerance for the presence of animals (e.g., Hindu) than others. Similarly, within a culture, some individual people have greater tolerance than others and we argue that understanding this dynamic is critical for implementing effective conservation policy.

If we accept the basic tenet that human-carnivore conflict is mediated by the competition for shared resources— be they space, prey, or domesticated animals— then, conceptually, it should be a relatively straightforward exercise to develop strategies to mitigate conflict. In essence, conflict prevention depends on (i) identifying ecological and social conditions that mediate interactions between wildlife and people (Treves et al. 2004), (ii) understanding how interactions can escalate into conflict, and (iii) developing effective outreach or intervention strategies to minimize the risk of future conflict. Ecologists and social scientists have been effective in identifying the ecological space where humans and wildlife are most likely to interact (e.g., Kretser et al. 2008, 2009) and what causes some interactions to escalate into conflict, but markedly less successful in integrating the two into forecasting tools. This of course leads to the question of do we really need to take an integrative approach to managing conflict? We suggest the answer to that question is yes— a holistic, integrative approach can be a powerful tool for managing the risk of conflict, particularly if the approach is spatially explicit to allow the prediction of when and where conflict is most likely to occur. However, in order to reach that goal, we first need to understand the limitations of current approaches.

The primary objective of this paper is to develop a framework for integrating ecological and sociological data for use in modeling the spatial distribution of the risk of human–carnivore conflict. The paper begins with a brief review of methods used to predict conflict. We then propose a novel approach for integrating ecological and sociological data into a predictive modeling framework. We illustrate this approach using a practical example based on conservation planning for black bears (*Ursus americanus*) in the southwestern United States.

Ecological Approaches to Predicting Risk of Conflict

We define ecological approaches to predicting the risk of human-carnivore conflict as those solely based on ecological analyses of factors that influence the occurrence of conflict. Generally, these approaches are spatially explicit and employ predictive modeling to correlate landscape attributes to the occurrence of conflict. The spatially explicit models are then often used to project the risk of conflict, given the composition and arrangement of landscape attributes, at a larger spatial scale. The value of this approach is threefold. First, the data are relatively easy to acquire. In the United States, most state agencies, and a few federal agencies (i.e., Wildlife Services, United States Fish and Wildlife Service), regularly collect geo-referenced reports of human-wildlife conflict, including damage, depredation, and adverse encounters. Second, remotely-sensed biophysical data are readily, and in most cases freely, available from a number of data aggregators and websites (e.g., United States Geological Survey Seamless Server). Third, the remotely-sensed data is typically updated on a regular basis. For example, the National Landcover Data Set, which provides information on land cover types in the United States, is updated at 5-yr intervals— this allows the predictive models to be easily updated as landscape composition and other attributes change.

Ecological approaches to predicting risk of human-carnivore conflict are common in the literature. For example, Michalski et al. (2006) used such an approach to predict felid–livestock conflict in Brazilian Amazonia.

The authors examined the ecological correlates of jaguar (*Panthera onca*) and puma (*Felis concolor*) predation on livestock by interviewing livestock managers to collect information on the spatial distribution of depredation events. They then related the occurrence of jaguar and puma depredation to an array of remotely-sensed landscape attribute variables as well as livestock grazing practices.

Using this approach, the authors found that patterns of depredation could be explained by a combination of landscape and livestock management variables such as proportion of forest area, distance to the nearest riparian corridor, annual calving peak and bovine herd size (Michalski et al. 2006).

A similar approach was employed by Treves et al. (2004, 2011) to predict the risk of wolf (*Canis lupus*)–livestock conflict in the Upper Midwest of the United States. The authors used data on wolf-killed livestock collected by state wildlife agencies to compare landscape attributes between affected (suffered at least 1 depredation event) and unaffected (no depredations reported) sites to determine the spatial distribution of risk. Similar to Michalski et al. (2006), Treves et al. (2004, 2011) found that risk of depredation was a function of the juxtaposition of high quality wolf habitat with areas of intense livestock grazing.

These efforts illustrate the utility of using a biophysical approach in predicting the risk of human-carnivore conflict. The value of this approach lies in the relative simplicity of incorporating human land uses, carnivore biology, and land cover simultaneously (i.e., Treves et al. 2011).

But distinctly missing from this approach is a measure of the sociological component of conflict, most notably the attitudes and perceptions of people. We maintain that integrating sociological data into the established ecological framework for predicting and modeling conflict could offer better conflict risk assessment.

Sociological Approaches to Predicting Risk of Conflict

Sociological research on wildlife conflict typically focuses on problem identification, formulation of mitigation strategies, and evaluation of the success of management actions (e.g., Ring 2008, Treves et al. 2006). For the latter two foci, identifying stakeholders and understanding their characteristics, values, attitudes, and acceptance of different management actions is critical. For example, a review by Vaske et al. (2006) revealed that most research published in *Human Dimensions of Wildlife,* a leading journal in the field, has focused on attitudes, beliefs, values, norms, and satisfactions (62%), as compared with behavior-related research (18%). So how does an understanding of attitudes and beliefs help resolve human–wildlife conflict?

Much of the sociological research relies on the analysis of survey data collected from stakeholders designed to elicit information on relevant attitudes and perceptions. This information can then be correlated with stakeholder behaviors and, if correlations are strong, used to indirectly predict future behavior (Manfredo 2008). Of course, when correlations are weak, only direct measures of behavior will be effective (McCleery et al. 2006). Conceptually, this is not so different from limitations of ecologically-based predictors of conflict. However, unlike ecological data, sociological data generally are not as readily available nor spatially explicit. For example, sociological data are not regularly collected along with conflict data, so collection often requires a rigorously designed survey. Nevertheless, there is growing acknowledgement that there is a need to focus conflict management solutions on humans as well as wildlife (Baruch-Mordo et al. 2009, 2011).

Researchers have examined social and attitudinal variables that seemingly influence a range of perceptions about actual human–carnivore interactions. In general, they've found that perceptions of future interactions are related to past experiences. Not surprisingly, individuals with negative experiences typically have less tolerance for future conflict (Coluccy et al. 2001, Heberlien and Ericsson 2005). Tolerance is also informed by how individuals use land, be it for recreation, agricultural production, or resource extraction. For example, Kellert et al. (1996) found that perceptions of carnivores (including black bears) were more negative for people who worked in natural resources extractive industries or lived in rural areas (Kellert et al. 1996). By contrast, Kaczensky et al. (2004) found that positive perceptions of bears and wolves were related to higher levels of education and more knowledge about those species. Likewise, Siemer and Decker (2003) found that nearly 90% of reported bear encounters in New York were positive and people living in the core bear habitat, and arguably more knowledgeable about bears, were more tolerant of hypothetical interactions with bears around their homes compared to those living outside of the core habitat. What these disparate findings indicate is that the perception of risk by individuals is highly variable and can differ relative to education, predominant land use, and personal experience.

Defining the Landscape of Conflict

Both ecological and sociological approaches have been used successfully to predict the risk of human–carnivore conflict. However, limitations exist for each approach that potentially compromises their efficacy. To overcome these limitations, we propose a novel heuristic framework that integrates ecological and sociological data to better predict and mitigate human–carnivore conflict.

We elucidate this concept using an example focused on conservation planning for black bears in the southwestern United States. In the Southwest, black bears are near the southern extent of their geographic range and subpopulations are vulnerable to isolation and localized extinction (Atwood et al. 2011). Black bears also come into conflict with humans, particularly during years of hard mast failure (LeCount 1982, Baruch-Mordo et al. 2008). Because of this, bears in the region can be viewed as existing at the conservation–conflict nexus, where conservation planning and conflict mitigation should intersect. Our approach, will use spatially explicit modeling to illustrate the landscape of conflict— a landscape where animals interact simultaneously within an ecological and sociological landscape. We will use these interactions to determine the spatial distribution of risk of conflict.

Study Area

We sampled the occurrence of black bears in the Patagonia, Huachuca, and Santa Rita mountains in southern Arizona (Figure 1). The three mountain ranges are adjacent to each other; the Santa Rita Mountains are the northernmost, while the Patagonia and Huachuca mountains extend approximately 31 km and 4 km, respectively, into Sonora, Mexico. In Sonora, the Patagonia Mountains are separated by 7 km of desert basin from the northern extent of the large (≈5396 km^2) Sierra Mariquita- Sierra de los Ajos mountain range complex.

As a result, the Patagonia and Huachuca mountains likely play an important role in maintaining trans-border connectivity between Arizona and Sonora, which is important because black bears in Mexico were classified as "endangered of extinction" in 1986. We projected the findings of our predictive models to the Tumacacori Highlands, and the Dragoon and Whetstone mountains, in addition to the sampled mountain ranges.

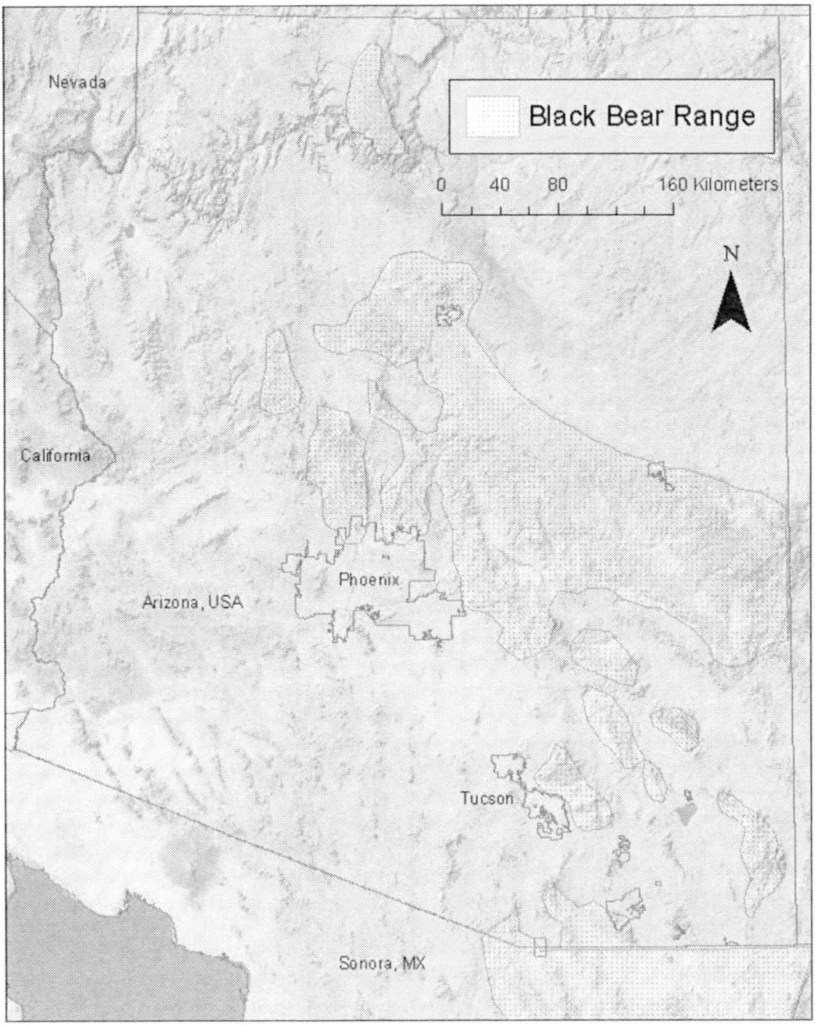

Figure 1. Black bear range in Arizona and the study area for investigating the utility of integrating ecological and human dimensions data for predicting risk of human-bear conflict.

Vegetation in the study area consisted of shrub and grassland associations at lower elevations, oak woodlands at mid-elevations, and Madrean evergreen woodlands at higher elevations (Brown, 1994; Bahre and Minnich, 2001). Predominant land use included livestock grazing and recreation. The area has experienced rapid urbanization over the last 20 years, characterized by a ≈20% increase in the human population and a ≈14% increase in housing density (http://quickfacts.census.gov/qfd/states/04000.html). The international boundary between Arizona and Sonora, Mexico, spans nearly 600 km, approximately 70% of which was fenced. The type of fence structure varied along the border (Figure 2 and 3), with some segments comprised of >4 m tall panels with either no openings or vertical gaps 5–10 cm wide and thus impermeable to most medium- and large-bodied mammals, while other sections consisted of barbed wire crossbar vehicle barriers (United States Customs and Border Protection, 2009) that were relatively permeable.

METHODS

Extensive details on our sampling methodology are available in Atwood et al. (2011). Briefly, we used non-invasive hair-snag corrals (Woods et al. 1999) deployed within 4 × 4 km grid cells to collect hair samples from black bears. Hair snag grids were deployed over three 10-14 day "capture" sessions in all 3 mountain ranges in the spring and summers of 2008 and 2009. Samples were retrieved from hair snags and submitted for genetic analyses to confirm species and determine individual identification.

We used point extraction and Euclidean distance routines in a 30-m resolution (i.e., 2006 USGS Seamless Server NED data) GIS to collect information on land cover and landscape covariates for hair-snag locations. We tested for collinearity among potential variables by examining tolerance and variance inflation factors (VIF) using weighted least squares regression, and excluded variables with tolerance scores <0.4 from analyses (Allison, 1999).

We then used the data generated from hair-snag sampling in program MARK (White and Burnham, 1999) to develop models of black bear occupancy relative to land cover (Madrean evergreen woodland [MEW], mixed conifer woodland [MXC], semi-desert grassland [DG], plains and Great Basin grassland [GBG], and oak woodland [OW]) and landscape covariates (slope [°], aspect, elevation [m], and distances to permanent water and roads [m]).

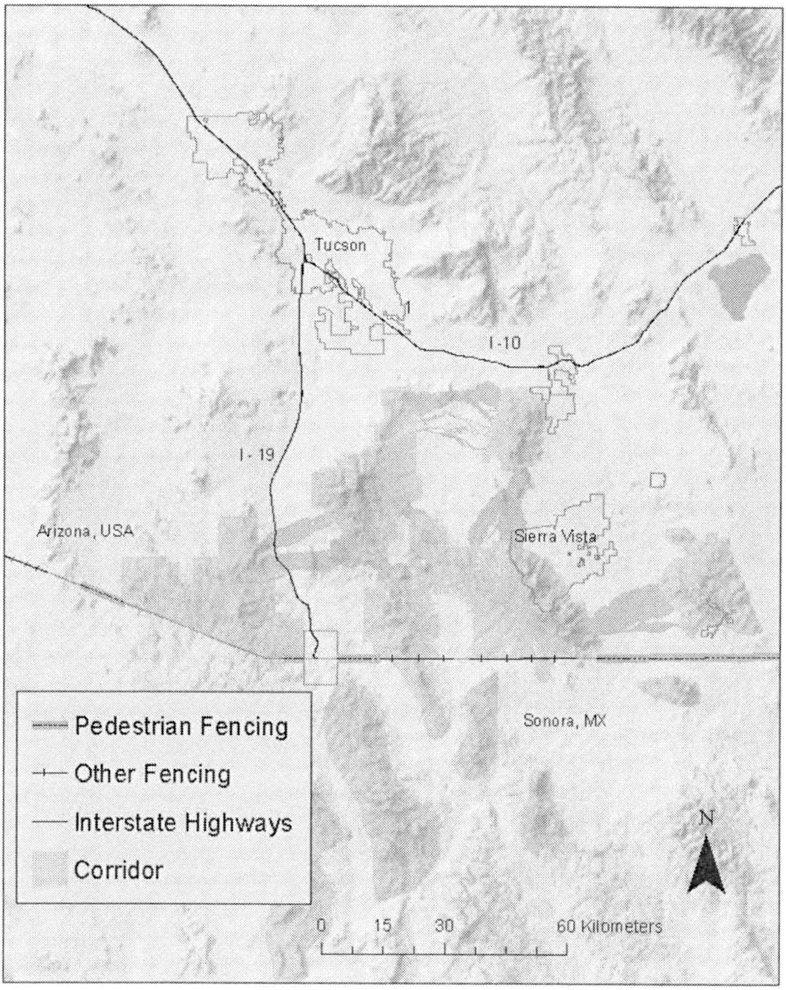

Figure 2. Corridor linkage created using the ecologically-based habitat suitability model and corresponding cost surface.

To frame this work in a conservation context, we used the habitat suitability and corridor models (i.e., ecological models) created by Atwood et al. (2011) to describe how bears used the landscape in the study area and moved between mountain ranges via movement corridors. We then integrated the simulated sociological data (described below) into the base models to develop the ecological-sociological models to project how negative human attitudes could affect habitat suitability and landscape connectivity.

Table 1. Grid layers (italics) and variables, reclassified grid cell values, and weighting factors used to assemble the ecologically-based and ecological-sociological habitat suitability models for the study area

Variable	Reclassified Cell Value	Weighting Factor Ecological Model	Weighting Factor Integrated Model
Landcover type		0.50	0.40
Madrean evergreen	100		
mixed conifer	68		
oak woodland	84		
semi-desert grassland	56		
Plains and Great Basin Grassland	1		
Distance to Water		0.35	0.30
<500m	25		
500-1000m	50		
>1000m	100		
Distance to Roads		0.05	0.04
>500m	25		
500-1250m	50		
>1250m	100		
Aspect		0.04	0.03
north	80		
east	35		
south	100		
west	25		
Elevation		0.03	0.02
>763m	20		
763-1219m	37		
1220-1981m	100		
1982-2591m	81		
2592-4000m	63		
Topographic Position		0.03	0.02
canyon bottom	50		
gentle slope	100		
ridge top	25		
Human Tolerance		not applicable	0.19
yes	100		
no	0		

Habitat and Corridor Modeling

We used the model-averaged occupancy values reported in Atwood et al. (2011) to create habitat suitability and corridor models. To develop the ecologically-based habitat suitability model (HSM), we reclassified the land cover grid by collapsing 35 landcover classes from the 2001 National Landcover Data (NLCD) set (e.g., Encinal oak woodland) into the land cover classes described above (e.g., MEW, MXC, DG, GBG, and OW), and assigned them a value from 0 (absolute non-habitat) to 100 (optimal habitat) based on detection probabilities scaled from occupancy models (Table 1). For the elevation, aspect, and distances to water and roads grids, we created 5, 4, 3, and 3 evenly-spaced bins, respectively, and assigned values (0–100) based on probabilities of occurrence at hair-snag stations (Table 1). To characterize topographic position, we used a moving window analysis in a GIS where we classified pixels as canyon bottom if the pixel elevation was at least 12 m less than the neighborhood average, a ridge-top if the pixel elevation was at least 12 m greater than the neighborhood average, a gentle slope if the pixel was neither a canyon bottom nor a ridge-top and had a slope <6°, and a steep slope if the pixel was neither a canyon bottom nor a ridge-top and had a slope >6°. The resulting topographic position index (TPI) grid was then reclassified using the method for the elevation grid but into three bins instead of 5. Finally, we combined the six individual grids using a weighted geometric mean algorithm (Table 1) where individual grid weighting factors were scaled to their proportional contribution based on the model-averaged Akaike weights.

For the integrated ecological-sociological modeling effort, we simulated human dimensions data by randomly assigning 1000 residential addresses in the study area pixel value scores of 0 (i.e., pixel is occupied by a person intolerant of large carnivores) or 100 (i.e., pixel is occupied by a person tolerant of large carnivores). We then used a moving window analysis (2 × 2 km window), similar to that used to characterize topographic position, to reclassify all pixels within the window to the same value as the focal pixel. We did this for 2 reasons. First, social scientists have documented a "neighborhood effect", where the magnitude of a decision or attitude for an agent (i.e., person) depends on the magnitudes of the decision or attitudes for neighboring agents (i.e., a community). In the context of human–carnivore conflict, a person with an a priori high tolerance for carnivores may lower their tolerance threshold if their neighbors have either a high vulnerability or low tolerance of conflict (Kretser et al. 2008). Second, much of the area is used for livestock grazing and the lower size limit of allotments and pastures is 4 km^2.

What this effort gave us was a spatially explicit layer of human attitudes, which we then combined with the 6 other individual grids using the same procedure for the ecologically-based HSM (Table 1).

To develop the corridor models, we converted the HSM (baseline and second run) into cost surfaces by calculating cell resistance (i.e., travel cost; cell resistance = 100 – pixel suitability) for each grid. The resulting cost surface grids were comprised of pixel values that reflected the cost of (or resistance to) movement through each individual grid cell, with increasing cell values representing increasing resistance to movement. We then applied a moving window analysis (200-m radius) to generate corridor models (pixel swaths; Atwood et al. 2011) that connected habitat cores while minimizing resistance to movement. We selected the best biological corridors (e.g., Bennett et al., 1994) based on the pixel swath that minimized within-swath gaps, maximized within-swath habitat suitability, and reduced edge effects by maintaining a minimum width equal to the radius of an estimated home range (LeCount 1982, Cunningham and Ballard 2004). All habitat and corridor modeling was done using the CorridorDesigner package for ArcGIS (Majka et al. 2007). To characterize the landscape of conflict, and examine how negative human attitudes influenced the distribution of conflict, we compared the spatial attributes of our ecological model predicting corridors with our socio-ecological model that also included the simulated sociological data. This provided insight into how adverse attitudes towards black bears could potentially impact conservation planning.

RESULTS

The ecologically-based habitat suitability model characterized 33% of the study area as relatively high quality habitat (≥60 suitability quantile). This habitat occurred mostly in the focal mountain ranges, so we used those as wildland blocks to connect via the corridor models. The integrated ecological-sociological suitability model characterized 24% of the study area as high quality habitat, most of which occurred in the focal mountain ranges. A comparison of the two HSM indicated that for the integrated model, habitat suitability in the Huachuca Mountains declined by 5%, followed by 3% and 1% declines in the Patagonia and Santa Rita mountains, respectively. Habitat quality also declined in the Tumacacori Highlands, and Dragoon and Whetstone mountains, but the declines were negligible (i.e., <1%). All of the declines in habitat suitability occurred in mid-elevation oak woodland habitat, which functioned as critical foraging habitat for black bears (LeCount 1982).

The ecologically-based cost surface yielded relatively high quality corridors (Figure 2). The length to narrowest width ratios for the corridors linking the mountain ranges was 6.8:1 (range: 1.2:1–12.1:1; SE = 1.11), with the highest quality corridor linking the Santa Rita Mountains and the Huachuca-Patagonia complex, followed by the corridors linking the Huachuca-Patagonia complex to the Dragoon Mountains, Whetstone Mountains, and the Tumacacori Highlands, and the Santa Rita Mountains to the Whetstone Mountains, respectively. All of these corridors contained >57% suitable habitat. By contrast, the integrated ecological-sociological based cost surface yielded substantially lower quality corridors (Figure 3), with length to narrowest width ratios averaging 47:1 (range: 13.2:1–102.1:1; SE = 4.11), and the highest quality corridor linking the Santa Rita Mountains and the Huachuca-Patagonia complex, followed by corridors linking the Huachuca-Patagonia complex to the Tumacacori Highlands, Dragoon Mountains, the Santa Rita Mountains to the Whetstone Mountains, and Huachuca-Patagonia complex to the Whetstone Mountains, respectively. All of these corridors contained <21% suitable habitat, rendering them biologically degraded compared to the corridors estimated from the ecologically-based cost surface.

DISCUSSION

Our study revealed important findings about the potential utility of integrating ecological and sociological data for use in predicting the spatial distribution of risk of conflict. First, if collected at the appropriate spatial scale (e.g., parcel ownership), it is relatively straightforward to create a spatially explicit projection of attitudes and perceptions. We demonstrated this using a novel approach where we integrated the simulated survey data on tolerance of large carnivores into habitat suitability and cost surface models. Second, our approach has heuristic value in the context of conservation planning, because it can be used to project how human attitudes and perceptions might be spatially distributed across a landscape. Third, and arguably most important, by integrating a simulated spatial layer into the HSM representing human tolerance towards large carnivores, we were able to depict how low tolerance of carnivores can potentially degrade the functional quality of otherwise highly suitable movement corridors. Given the above, we believe our approach has merit for future research and guiding efforts aimed at mitigating risk of human-carnivore conflict, particularly if used at the conservation planning stage.

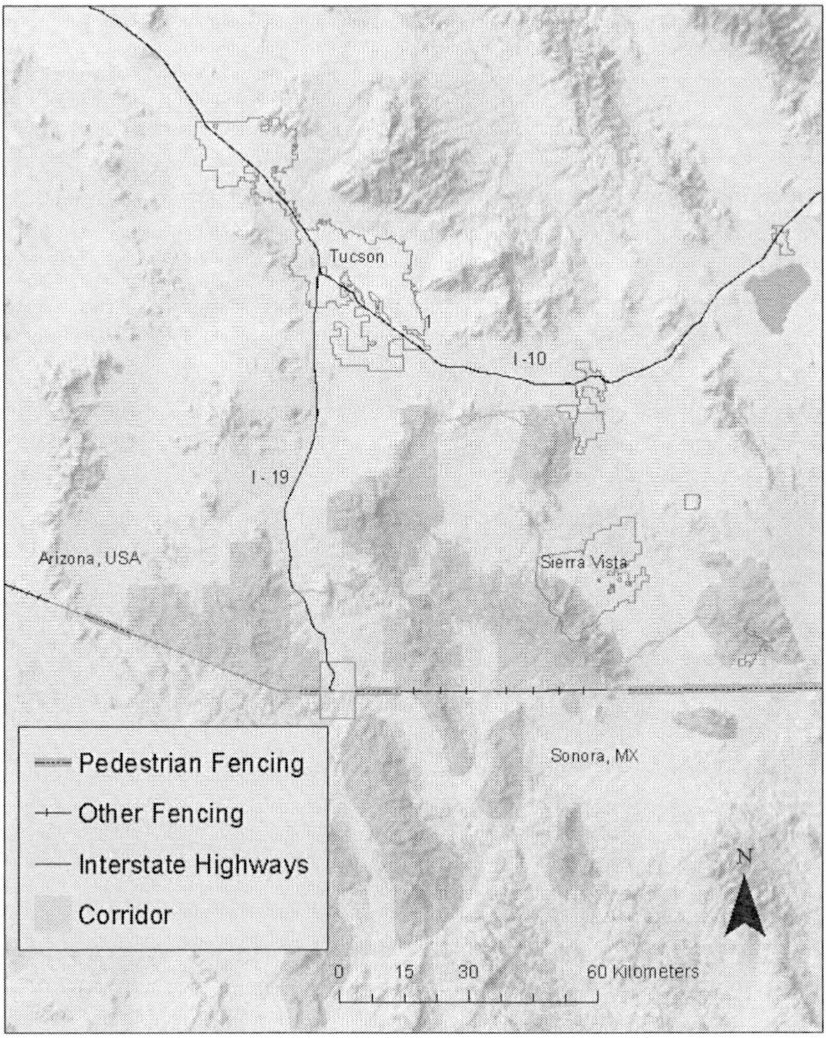

Figure 3. Corridor linkage created using the integrated ecological-human dimensions habitat suitability model and corresponding cost surface.

The ecological determinants of conflict tend to operate at a fine scale, whereas trends in human attitudes are typically only made available at a more coarse scale. The mismatch of the spatial scales at which the two processes occur has been a fundamental impediment to the integration of ecological and sociological data. As ecologists know well, no question framed in a spatial context can be addressed without explicitly identifying the resolution at which observations are collected or projected. Indeed, patterns observed on one scale may not be apparent on another scale (Guisan and Thuiller 2005), so acknowledging that scale influences the nature, distribution, and interpretation of interactions between those processes is critical (Cumming et al. 2006). That said, issues of privacy often preclude reporting fine-scale sociological data, and that has limited the efficacy of previous attempts to integrate sociological and ecological data. The approach we employed, using simulated sociological data, was a viable alternative for reconciling the concern over reporting sensitive information while still displaying the data at a meaningful spatial scale. Using a moving window analysis that explicitly incorporated a neighborhood effect, we were able to use individual point location data to project to a neighborhood scale and thereby avoid concerns over displaying spatially identifiable personal information.

Integrating ecological and sociological data in a spatial modeling framework has myriad applications. We demonstrated the heuristic value of using the framework to address an applied conservation issue centered on reconciling connectivity planning with conflict mitigation. Conservation strategies for at-risk species have been developed using models of varying complexities, including population models (e.g. population viability analyses), landscape models (e.g. resource selection functions), and spatially explicit dynamic models [e.g. spatially explicit, individual-based model (SE-IBM)] (Shenk and Franklin 2001, Wiegand et al. 2004), but we are unaware of any spatially explicit ecological models that also include sociological data. For large carnivores distributed in small subpopulations, such as desert black bears, the main factors causing localized extinction are loss or conversion of their habitat and increased illegal killing by humans in response to conflict (Ferreras et al. 2001). Obviously, both create controversies and challenges for the conservation of large carnivores. Black bears, for example, come into conflict with humans mainly through competition over food resources, primarily crops and refuse, but also occasionally neonatal livestock (Baruch-Mordo et al. 2008, LeCount 1982). Hence, there are often competing pressures on wildlife managers to mitigate conflict while also maintaining viable populations. What is clearly needed then, is a tool that allows wildlife managers and land use planners to identify areas of high biological suitability that occur in proximity to areas of high human tolerance. This information can then be used to prioritize mitigation efforts appropriately, while also minimizing the ecological and economic costs of trial and error for at-risk species. We believe our integrated modeling approach holds promise in that regard.

Ecological factors are the primary drivers of the spatial distribution of high quality habitat and movement corridors. However, it is important to note that when high quality habitat and corridors occur in areas occupied by intolerant humans, illegal killing or harassment can functionally degrade the conservation value of those areas.

Our modeling efforts support that it is important to know the spatial distribution of tolerance before extensive resources are invested into implementing conservation plans such as purchasing tracts of land or entering into easement agreements. When we integrated the sociological layer into the ecological HSM and subsequent cost grid, we saw a marked decrease in the quality of movement corridors, such that the length:width of corridors increased 7-fold. The result of this is long, narrow corridors that contain less suitable habitat and restrict movement between wildland blocks to a fine corridor swath. Large carnivores, in general, have large area requirements—even when using movement corridors.

As a result, corridors that become too narrow no longer offer refugia from humans or sympatric carnivores, and thus lose their biological integrity (Beier et al. 2008, Chetkiewicz and Boyce 2009). From a conservation planning perspective, it is vitally important to be able to predict where on the landscape habitat suitability is likely to interact with human attitudes to determine functional habitat suitability. That information can then be used to not only identify the best biological habitat and corridors, but also the most socially acceptable.

The dearth of economic resources available to conserve or recover carnivore populations necessitates the development or refinement of methods to identify conservation priorities (Margules and Pressey 2000). Our integrated approach represents a novel and useful tool for the conservation planner's toolbox. By developing a spatially explicit modeling approach that integrates ecological and sociological data, we created a predictive modeling framework that is flexible to changes in attitudes and landscape characteristics, avoids concerns over the disclosure of sensitive private information, and allows users to balance biological and societal concerns when setting planning priorities.

The information we present here, if incorporated into carnivore management plans, may also aid in ameliorating the adverse effects of conflict with humans, which is critical to the long-term societal acceptance of large carnivores on the landscape.

References

Atwood, T.C., J.K. Young, J.P. Beckmann, S.W. Breck, J. Fike, O.E. Rhodes, Jr., and K.D. Bristow. 2011. Modeling connectivity of black bears in a desert sky island archipelago. *Biological Conservation* 144:2851-2862.

Allison, P.D., 1999. *Logistic Regression using the SAS System: Theory and Application.* SAS Institute Inc., Cary, North Carolina, USA.

Bahre, C.J., and R.A. Minnich. 2001. Madrean oak woodlands along the Arizona/Sonora boundary. *Desert Plants* 17, 3–14.

Baruch-Mordo, S., S.W. Breck, K.R. Wilson, and D.M. Theobald. 2008. Spatiotemporal distribution of black bear–human conflicts in Colorado, USA. *Journal of Wildlife Management* 72:1853-1862.

Baruch-Mordo, S., S.W. Breck, K.R. Wilson, and J. Broderick. 2009. A tool box half full: how social science can help solve human-wildlife conflict. *Human Dimensions of Wildlife* 14: 219–223.

Baruch-Mordo, S., S.W. Breck, K.R. Wilson, and J. Broderick. 2011. The Carrot or the Stick? Evaluation of Education and Enforcement as Management Tools for Human-Wildlife Conflicts. *PLoS ONE* 6: e15681. doi:10.1371/journal.pone.0015681

Beier, P., D.R. Majka, and W.D. Spencer. 2008. Forks in the road: choices in procedures for designing wildland linkages. *Conservation Biology* 22: 836-851.

Bennett, A.F., K. Henein, and G. Merriam. 1994. Corridor use and elements of corridor quality: chipmunks and fencerows in a farmland mosaic. *Biological Conservation* 68:155–165.

Brown, D.E., 1994. *Biotic Communities: Southwestern United States and Northwestern Mexico*. University of Utah, Salt Lake City, UT, USA.

Chetkiewicz, C-L.B. and M.S. Boyce. 2009. Use of resource selection functions to identify conservation corridors. *Journal of Animal Ecology* 46:1036-1047.

Coluccy, J.M., R.D. Drobney, D.A. Graber, S.L. Sheriff, and D.J. Witter. 2001. *Attitudes of central Missouri residents toward local giant Canada geese and management alternatives.* Wildlife Society Bulletin 29: 116–123.

Cumming, G.S., D.H.M. Cumming, and C.L. Redman. 2006. Scale mismatches in social-ecological systems: causes, consequences, and solutions. *Ecology and Society* 11:14.

Cunningham, S.C. and W.B. Ballard. 2004. Effects of wildfire on black bear demographics in central Arizona. *Wildlife Society Bulletin* 32, 928–937.

Dalerum, F., M.J. Somers, K.E. Kunkel, and E.Z. Cameron. 2008. The potential for large carnivores to act as biodiversity surrogates in southern Africa. *Biodiversity Conservation* 17:2939-2949.

Ferreras, P., P. Gaona, F. Palomares, and M. Delibes. 2001. Restore habitat or reduce mortality? Implications from a population viability analysis of the Iberian lynx. *Animal Conservation* 4: 265–274.

Guisan, A. and W. Thuiller. 2005. Predicting species distribution: offering more than simple habitat models. *Ecology Letters* 8:993-1009.

Heberlein, T.A. and G. Ericsson. 2005. Ties to the countryside: Accounting of urbanites attitudes toward hunting, wolves, and wildlife. *Human Dimensions of Wildlife* 10:213–227.

Kaczensky, P., M. Blazic, and H. Gossow. 2004. Public attitudes towards brown bears (*Ursus acrtos*) in Slovenia. *Biological Conservation* 118:661–674.

Kellert, S.R., M. Black, C.R. Rush, and A.J. Bath. 1996. Human culture and large carnivore conservation in North America. *Conservation Biology* 10:977–990.

Kretser, H.E., P.J. Sullivan, and B.A. Knuth. 2008. Housing density as an indicator of spatial patterns of reported human–wildlife interactions in northern New York. *Landscape and Urban Planning* 84:282-292.

Kretser, H.E., P.D. Curtis, and B.A. Knuth. 2009. Landscape, social, and spatial perceptions of human-black bear interactions in the Adirondack Park, NY. *Human Dimensions of Wildlife* 14:393-406.

LeCount, A.L. 1982. Population characteristics of Arizona black bears. *Journal of Wildlife Management* 46, 861–868.

Majka, D., J. Jenness, J., and P. Beier. 2007. *CorridorDesigner: ArcGIS tools for Designing and Evaluating Corridors*. <http://corridordesign.org>.

Manfredo, M.J. 2008. Who cares about wildlife? *Social science concepts for exploring human-wildlife relationships and conservation issues.* New York: Springer.

Margules, C. R. and R. L. Pressey. 2000. Systematic conservation planning. *Nature* 405:243-253.

McCleery, R.A., R.B. Ditton, J. Sell, and R.R. Lopez. 2006. Understanding and improving attitudinal research in wildlife sciences. *Wildlife Society Bulletin,* 34, 537–541.

Michalski, F., R.L.P. Boulhosa, A. Faria, and C.A. Peres. 2006. Human-wildlife conflicts in a fragmented Amazonian forest landscape: determinants of large felid depredation on livestock. *Animal Conservation* 9:179-188.

Ring, I. 2008. Towards a framework for integrating human dimensions in wildlife management. In M.J. Manfredo, J.J. Vaske, P.J. Brown, D.J. Decker, and E.A. Duke (Eds.), *Wildlife and society: The science of human dimensions* (pp. 90–103). Washington, DC: Island Press.

Shenk, T.M. and A.B. Franklin. 2001. *Modeling in Natural Resource Management: Development, Interpretation, and Application.* Island Press, Washington.

Siemer, W.F. and D.J. Decker. 2003. 2002 New York State Black Bear Management Survey: Study Overview and Findings Highlights. *Human Dimensions Research Unit #03-06*, Department of Natural Resources. New York State College of Agriculture and Life Sciences, Cornell University, Ithaca, New York.

Sillero-Zubiri, C., and M.K. Laurensen. 2001. Interactions between carnivores and local communities: conflict or co-existence? In: J. Gittleman, K. Funk, D. Macdonald, and R. Wayne (Eds.), *Carnivore Conservation, Conservation Biology Series 5.* Cambridge University Press, Cambridge, pp. 282–312.

Thirgood, S.J., S.M. Redpath, I. Newton, and P. Hudson. 2000. Raptors and red grouse: conservation conflicts and management solutions. *Conservation Biology* 14, 95–104.

Treves, A., L. Naughton-Treves, E.K. Harper, D.J. Mladenoff, R.A. Rose, T.D. A. Sickley, and A.P. Wydeven. 2004. Predicting human–carnivore conflict: a spatial model derived from 25 years of data on wolf predation on livestock. *Conservation Biology* 18:114-125.

Treves, A., R. Wallace, L. Naughton-Treves, and A. Morales. 2006. Co-managing human–wildlife conflicts: A review. *Human Dimensions of Wildlife* 11:383–396.

Treves, A., K.A. Martin, A.P. Wydeven, and J.A. Wiedenhoeft. 2011. Forecasting environmental hazards and the application of risk maps to predator attacks on livestock. *BioScience* 61:451-458.

United States Customs and Border Protection, 2009. *Fence Specifications.* http://www.cbp.gov/xp/cgov/border_security/ti/ti_mr/ctimr/about_ctimr/ctimr_work/fence/fence_specs/.

Vaske, J.J., L.B. Shelby, and M.J. Manfredo. 2006. Bibliometric reflections on the first decade of Human Dimensions of Wildlife. *Human Dimensions of Wildlife* 11:79–87.

Woods, J.G., D. Paetkau, D. Lewis, B.N. McLellan, M. Proctor, and C. Strobeck. 1999. *Genetic tagging of free-ranging black and brown bears.* Wildlife Society Bulletin 27, 616–627.

White, G.C. and K.P. Burnham. 1999. Program MARK: survival estimation from populations of marked animals. *Bird Study* 46:S1, S120-S139.

Wiegand, T., F. Knauer, P. Kaczensky, and J. Naves. 2004. Expansion of brown bears (*Ursus arctos*) into the Eastern Alps: a spatially explicit population model. *Biodiversity and Conservation,* 13, 79–114.

In: Carnivores
Editors: F. I. Álvares and G. E. Mata

ISBN 978-1-62257-323-3

Chapter 5

MODELING DISJUNCT CARNIVORE DISTRIBUTIONS: THE CASE OF THE WOLF (*CANIS LUPUS*) IN THE IBERIAN PENINSULA

A. Márcia Barbosa[*1,2]**, *Miguel A. Puerto***[3]**, *Diogo Figueiredo***[1]** *and Raimundo Real***[4]

[1]"Rui Nabeiro" Biodiversity Chair, CIBIO - University of Évora, Évora, Portugal
[2]Division of Biology, Imperial College London, Silwood Park Campus, Ascot (Berkshire), United Kingdom
[3]Instituto Español de Oceanografía, Centro Oceanográfico de Málaga, Muelle Pesquero s/n., Fuengirola, Spain
[4]Biogeography, Diversity and Conservation Lab, Department of Animal Biology, Faculty of Sciences, University of Málaga, Málaga, Spain

ABSTRACT

The gray wolf (Canis lupus) is a generalist species whose distribution was originally the widest among wild carnivores but suffered a marked regression due to human persecution during the 19th and 20th centuries.

* Corresponding author. E-mail:. Address: "Rui Nabeiro" Biodiversity Chair, CIBIO - University of Évora, Largo dos Colegiais, 7004-516 Évora, Portugal. Telephone: (+351) 266 740 800.

Legal protection after 1970 allowed wolves to recover significant parts of their range, including, in some cases, heavily humanized and modified landscapes. Nowadays, many wolf populations keep expanding, although the viability of many others is still at risk due to small population sizes and loss of genetic diversity. The Iberian Peninsula holds Western Europe's largest wolf population, which is currently divided into two distinct nuclei: a large and practically continuous one spreading through most of the northern half of this region, and a smaller and subdivided one in a southern mountainous area. A distribution model based on the whole Iberian wolf distribution overlooked the southern nucleus, suggesting biogeographic differences between them. This can happen due to local adaptation events, and points to the need for caution when modelling species or populations with disjunct distributions. When we modelled both wolf nuclei separately, the southern nucleus showed indeed different environmental and biogeographical properties, with a trend towards a metapopulational structure. We discuss the implications of these models for the conservation and management of wolves in the Iberian Peninsula and of other species facing similar biogeographic situations elsewhere.

INTRODUCTION

The wolf (*Canis lupus* L., 1758) is a generalist and highly adaptable carnivore that was once widely spread across the northern hemisphere, which made it the wild mammal with the broadest natural distribution. Yet, human persecution during the nineteenth and twentieth centuries eliminated it from much of its original range. By the mid of the twentieth century, wolf distribution was basically reduced to the northern half of North America, northern and central Asia, eastern Europe, southern Scandinavia and the Mediterranean peninsulas, and wolf populations were mostly relegated to remote undisturbed areas. However, worldwide legal protection in the early 1970s allowed the species to recolonize a significant part of its former distribution range, including highly altered and humanized landscapes (Mech 1995, Route and Aylsworth 1999, Boitani 2000). Nowadays, many wolf populations are expanding significantly, although the viability of many others is still at risk due to low population sizes and loss of genetic variability (Ellegren 1999, Vila et al. 2003).

In the Iberian Peninsula (SW Europe), wolf regression became patent during the twentieth century. Wolf populations reached their minimum in the 1970s, but later Blanco et al. (1990, 1992) and Arberas et al. (1999) found that many Iberian wolf populations were recovering rather than heading for extinction. Two main distribution nuclei currently occur in the Peninsula, one occupying a large and, at this scale, virtually continuous area in the northwest, including N Portugal and NW Spain; and the other occurring in a relatively small area and divided in two separate populations in Sierra Morena, S Spain (Figure 1). Although the Iberian wolf population is the largest in Western Europe (IUCN, 2010), the wolf is currently considered endangered in Portugal (Cabral et al., 2005) and nearly threatened in Spain (Palomo and Gisbert, 2002). The Iberian wolf population suffered a recent bottleneck and currently presents low levels of heterozygosity and a distinctive genetic structure (vonHoldt et al., 2011) with some unique mitochondrial haplotypes (Pilot et al., 2010). Although it is not consensual, some authors consider that Iberian wolves constitute a separate subspecies (*C. l. signatus* Cabrera, 1907; Palomo and Gisbert, 2002).

As a large carnivore, the wolf plays an important role at the top of food chains, so helping maintain the equilibrium of ecosystems. The disappearance of wolves, as well as other large carnivores, from areas where they originally occurred has remarkable consequences on prey community structure and on subtle but important ecological patterns involving interspecific behavioral relationships (Berger 1999). On the other hand, wolf recovery has boosted up old conflicts with humans, generated mainly by predation on domestic livestock and on wild ungulates where they are an economically important hunting resource (Blanco et al. 1992, Fico et al. 1993, Mech 1995, Treves et al. 2004). Management plans are therefore needed to allow for the conservation of wolves, protecting their vulnerable populations, whilst minimizing the damages caused by expanding, well-established populations.

Corsi et al. (1999) defended a landscape approach as the most adequate for the management of this wide-ranging species. National species surveys, whether or not included in species distribution atlases, are a most useful basis for such studies, as they provide comprehensive data suitable for analyzing the distribution of species on a wide scale. Large-scale distribution models can be applied even when country-wide information is limited (Corsi et al. 1999), and extrapolated to predict potential distribution of species on finer resolution scales (Barbosa et al. 2003, 2010, Araújo et al. 2005). They can also forecast the evolution of species' distribution ranges, their responses to environmental change, and the probability of persistence of different populations (Araújo and Williams 2000, Real et al. 2009). A biogeographical perspective may also be important in the assessment of the conservation status of disjunct population nuclei (Acevedo and Real 2011).

Here we model wolf distribution in the Iberian Peninsula, investigating whether one model can account for both distribution nuclei or if separate models are necessary. We also compare the factors that describe the areas occupied by each nucleus, and discuss the implications of these models for the conservation and management of the wolf and other large carnivores alike, in the study area as well as in other parts of its extensive range.

METHODS

The study area was the Iberian Peninsula, at the south-western edge of Europe (Figure 1). This is a nearly 600,000-km^2 heterogeneous region comprising the mainland territories of Portugal and Spain, and linked to the European continent by a narrow and mountainous isthmus. It thus constitutes a discrete biogeographical unit appropriate for studies on species distributions (e. g. Real et al. 2009, Barbosa et al. 2010) – especially so in this case, as the Iberian wolf population is genetically distinct from the remaining European populations (vonHoldt et al. 2011).

Distribution data consisted of Iberian wolf presences and absences on UTM 10×10-km grid cells (Figure 1). Data from Portugal were taken from I. C. N. (1997) and are the result of a national wolf survey held in 1997 and previous years. Data for Spain were taken from a mammal distribution atlas (Palomo and Gisbert 2002) which gathered distribution records of various sources from a roughly similar time period. We modeled distribution data from the Iberian Peninsula altogether and from the two disjunct wolf nuclei separately to see whether a single model could account for the whole peninsular distribution of this species or, on the contrary, if two different models were needed. For the modeling of each nucleus, presences of wolves from the other nucleus were treated as absences.

Predictor variables, representative of physiography, climate and human activity (Table 1), were digitized and interpolated by Barbosa et al. (2003, 2009). We corrected the values of solar radiation according to Barbosa et al. (2011). The variables were chosen on the basis of availability at this scale and relation to factors that may influence wolf distribution more directly. Our aim was not to create explanatory models with the variables directly conditioning wolf distribution, but rather to predict its potential distribution on the basis of available variables, which are assumed to be correlated with more proximate causal factors (see Austin 2002, Robertson et al. 2003). Models that rely on indirect links between environmental variables and species presence can predict distribution at least as well as mechanistic models built on variables more directly related to species physiology (Austin 2002, Robertson et al. 2003).

Table 1. Variables used to model the distribution of *Canis lupus*, in alphabetical order of their code

Code	Variable
aet	Mean annual actual evapotranspiration (mm) (=minimum [*PET, Prec*])
alti	Mean altitude (m) [(1)]
altr	Altitudinal range (m)
d100	Distance to the nearest town with more than 100,000 inhabitants (km) [(3)]
d500	Distance to the nearest town with more than 500,000 inhabitants (km) [(3)]
dfro	Mean annual number of frost days (minimum temperature $\leq 0^{\circ}$C)[(2)]
dhi	Distance to the nearest highway (km) [(3)]
dpre	Mean annual number of days with precipitation $\geq$ 0,1 mm [(2)]
hjan	Mean relative air humidity in January at 07:00 hours (%) [(2)]
hjul	Mean relative air humidity in July at 07:00 hours (%) [(2)]
inso	Mean annual insolation (hours/year) [(2)]
lati	Latitude (degrees N) [(3)]
long	Longitude (degrees E) [(3)]
mp24	Maximum precipitation in 24 hours (mm) [(2)]
pet	Mean annual potential evapotranspiration (mm) [(2)]
prec	Mean annual precipitation (mm) [(2)]
rmp	Relative maximum precipitation *(=MP24/Prec)*
slop	Mean slope (degrees) (calculated from *Alti*)
srad	Mean annual solar radiation (kwh/m^2/day) [(2)]
temp	Mean annual temperature (°C) [(2)]
tjan	Mean temperature in January (°C) [(2)]
tjul	Mean temperature in July (°C) [(2)]
vhum	Annual relative air humidity range (%) *(=|HJan-HJul|)*
vtem	Annual temperature range (°C) *(=TJul-TJan)*

Sources:

[(1)] U. S. Geological Survey (1996);

[(2)] Font (1983, 2000);

[(3)] I. G. N. (1999); data on the number of inhabitants of urban centers taken from from Enciclopédia Universal (http://www.universal.pt) for Portugal and from the Instituto Nacional de Estadística (http://www.ine.es) for Spain, both in 1999.

The UTM 10×10 km grid and the limits of the study area were downloaded from the EDIT Geoplatform (Sastre et al. 2009). We used Quantum GIS 1.7 (Quantum GIS Development Team, 2010) and its GRASS plugin (Geographic Resources Analysis Support System; GRASS Development Team 2010) to clip the grids with the limits of the study area. The values of the variables were averaged over the UTM cells with the *raster* package in R 2.11 (R Development Core Team 2010).

To avoid a spurious effect of surface area on the probability of the species being present, only complete UTM cells, and not those that were cut by study area borders or the unions between UTM zones, were used for model building. This left a total number of 5464 analyzed UTM cells, 1239 of which (22.7%) had wolf presence records. The northern nucleus had 1185 presence cells (corresponding to a 21.7% prevalence in the whole study area), while the southern nucleus had the remaining 54 cells (1.0% prevalence). The models were then applied to the whole study area of 6180 cells for which average values of the variables could be obtained.

We modeled wolf presences and absences using the favorability function described by Real et al. (2006), which is a generalized linear model that corrects the effect of the uneven proportion of presences and absences in the analyzed sample. In this way, output model values strictly reflect environmental favorability for the species or population, independently of its prevalence in the study area, thus enabling direct between-model comparison even among species or populations with very different prevalence (Real et al. 2006; Jiménez-Valverde and Lobo 2006). This function has been successfully used to build comparable models for different species (Real et al. 2009, Barbosa and Real 2012), for parts of the distribution of the same species in different countries (Barbosa et al. 2009) and for different subspecies within the same species (Acevedo and Real 2011), for example.

Statistical theory predicts an increase in false findings when a large number of variables is analysed, due to the increase in type I errors under repeated testing. García (2003) recommended controlling type I errors in ecological research by evaluating the false discovery rate (FDR). We controlled this error using the procedure for all forms of dependency among test statistics (Benjamini and Yekutieli, 2001), and only accepted variables under a FDR value lower than 0.05.

Variables were then included in the models using a stepwise procedure (Araújo et al., 2005; Bulluck et al., 2006; Arntzen and Espregueira-Themudo, 2008) using SPSS Statistics 17.0 and the forward conditional option. Stepwise selection is a useful and effective tool to infer distribution patterns inductively from observed data, when no theory or previous hypotheses exist about the importance of each variable (Guisan and Zimmermann, 2000; Real et al., 2009).

The capacity of each model to discriminate between the wolf presences and absences was assessed with the Area Under the receiver operating characteristic (ROC) Curve (AUC). This is a widely used model evaluation measure that provides a single-number discrimination measure across all possible classification thresholds for each model, thus avoiding the subjective selection of one threshold (Fielding and Bell 1997).

We remark that, as any discrimination measure, the AUC depends on thresholds (just not on one particular threshold) to convert continuous model predictions into binary classifications, and it is conditioned by species prevalence (e. g., Lobo et al. 2008).

We thus performed also the Hosmer and Lemeshow goodness-of-fit test to assess the calibration of the models, i. e., the reliability of model predictions (Liu et al. 2009).

RESULTS

The predictions obtained from the total Iberian wolf data reflected mainly the distribution of the (substantially larger) northern nucleus, overlooking the southern one (Figure 1a). A single model was thus unsuitable to account for the whole Iberian wolf distribution. This was not due to the inclusion of spatial variables (latitude and longitude) in the models, as excluding these variables (Table 2) produced a highly similar model that still overlooked the southern nucleus (Figure 2a). The distributions of the two spatially segregated wolf nuclei thus needed to be modeled separately.

All models included variables related to topography, continentality, environmental energy and water availability. The total and the northern Iberian wolf models included also human activity and spatial situation, with a trend for more favorable areas towards the north and the west (Table 2).

When modeling was repeated without the spatial variables for the Iberian and the Northern Iberian populations, the obtained potential distribution maps were highly similar (Figure 2) and predictions were highly correlated with those of the models with spatial variables (Pearson's $r = 0.919$ for both the Iberian population and the northern nucleus, $p \ll 0.001$).

All models attained high discrimination capacity as given by AUC, whereas model calibration varied more widely (Table 3).

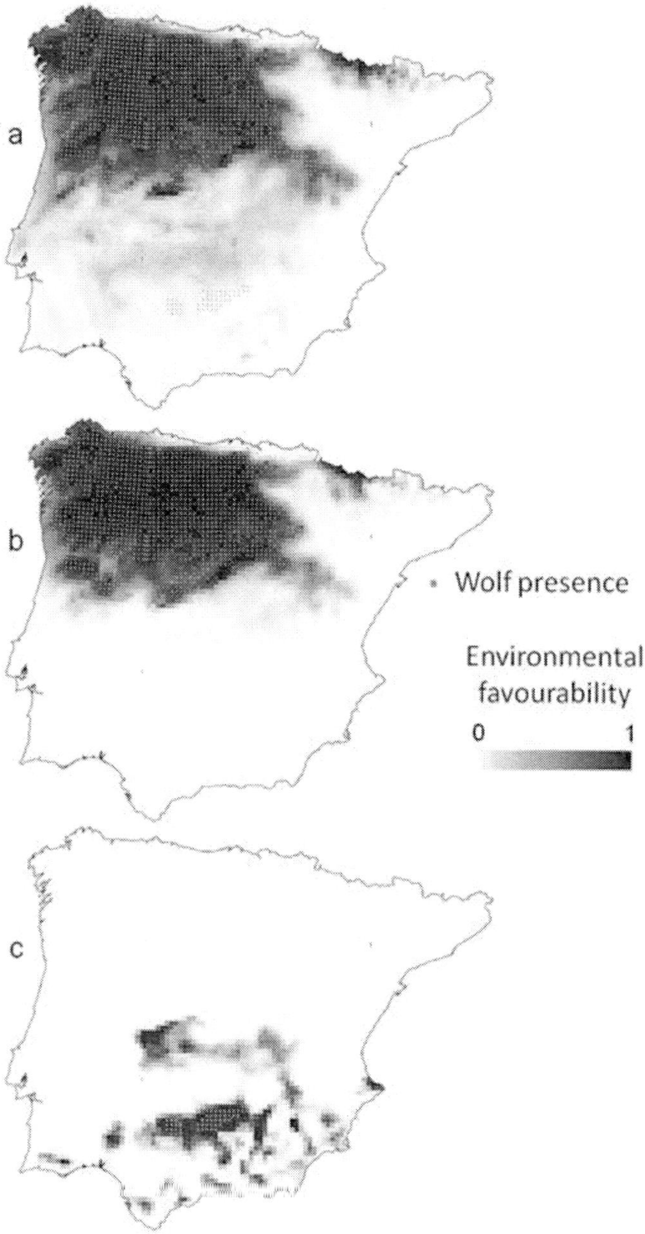

Figure 1. Wolf (*Canis lupus)* occurrence on 10x10-km UTM grid cells [data from I. C. N. (1997) for Portugal and Palomo and Gisbert (2002) for Spain] and the favorability models obtained for the whole Iberian distribution (a) and for the northern (b) and southern nuclei (c) separately.

Table 2. Variables (and their coefficient estimates in parentheses) selected to predict the distribution of the Iberian wolf and the distributions of the northern and southern nuclei separately. Variables are grouped according to the factors they are associated with. Variable codes as in Table 1. *Models excluding spatial variables

Factor	Iberia	North	South	Iberia*	North*
Intercept	(-63.332)	(-126.998)	(-267.755)	(-11.545)	(-8.368)
Topography	alti (0.002)	alti (0.002)			alti (-0.001)
					altr (0.001)
	slop (0.057)				
Energy			aet (0.025)	aet (-0.001)	aet (-0.002)
	pet (0.007)	pet (-0.016)	pet (0.044)		pet (-0.012)
	inso (-0.002)			inso (-0.001)	
	temp (-0.426)	temp (-0.327)		temp (-0.614)	temp (-0.636)
					tjan (0.127)
	tjul (0.325)	tjul (0.291)			tjul (-0.265)
		srad (0.744)	srad (37.055)		
Continentality	dfro (0.041)	dfro (0.037)	dfro (0.072)		
			vtem (1.81)	vtem (-0.128)	
Water	dpre (-0.029)	dpre (-0.037)	dpre (-0.124)		dpre (0.01)
	hjan (0.173)	hjan (0.158)		hjan (0.422)	hjan (0.349)
				hjul (-0.144)	
	vhum (-0.086)			vhum (-0.173)	
Location	lati (0.979)	lati (2.733)			
	long (-0.959)	long (-1.188)			
Disturbance			pm24 (0.025)		pm24 (-0.008)
				rmp (-10.082)	rmp (-9.74)
Human	d500 (0.004)	d500 (-0.007)		d500 (0.013)	d500 (0.01)
			dhi (-0.052)		

Table 3. Evaluation measures of the of total, northern and southern Iberian wolf distribution models

Model	Area Under the Curve (AUC)	Hosmer and Lemeshow calibration statistic (H-L)
Iberia	0.948	221.101, $p < 0.001$
North	0.978	18.127, $p = 0.020$
South	0.993	52.036, $p < 0.001$
Iberia*	0.931	207.503, $p < 0.001$
North*	0.959	28.820, $p < 0.001$

*Models excluding spatial variables; *p*: significance.

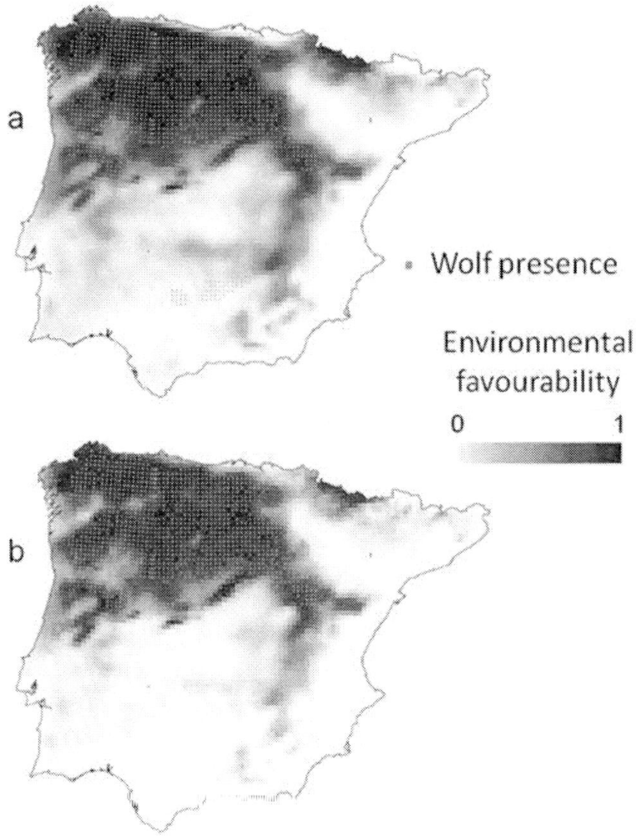

Figure 2. Wolf (Canis lupus) occurrence on 10x10-km UTM grid cells [data from I. C. N. (1997) for Portugal and Palomo and Gisbert (2002) for Spain] and the favorability models obtained for the whole Iberian distribution (a) and for the northern nucleus (b) without spatial variables.

The distribution of the northern wolf nucleus could also be well predicted by the overall Iberian model (AUC = 0.971 for the complete Iberian model, AUC = 0.956 for the Iberian model without spatial variables). However, the same did not occur for the southern wolf nucleus, with the Iberian model completely failing to predict its presence (AUC = 0.335 for the complete Iberian model, AUC = 0.285 for the Iberian model without spatial variables, indicating predictions were even worse than random).

Discussion

Our results show that the two main Iberian wolf nuclei respond to different environmental factors. The environmental favorability maps obtained for the two nuclei were clearly different, and the Iberian model was manifestly incapable of predicting the distribution of the southern nucleus (Figure 1).

The fact that the Iberian model mainly predicted the distribution of the northern nucleus is probably due to the much greater weight of this nucleus within the data set, given the large difference in the numbers of presences of both nuclei (Figure 1). It is important to remark that the incapacity of the Iberian model to predict the southern nucleus was not due to the inclusion of spatial variables in the analysis. These variables might indeed have restricted favorable predictions to the northern region, which contained most of the analysed occurrences; however, repeating the modeling procedure without using any spatial variables just produced very similar models (Figure 2), showing that it was not the spatial variables that were restricting the Iberian model to the northern half of the Peninsula.

Blanco and Cortés (2002) defended that wolf distribution is conditioned mainly by human activity and not by environmental factors, which would be demonstrated by this species' wide overall distribution. However, even a tolerant species is affected by environmental conditions, especially when human pressure restricts it to more adverse regions, where previously non-limiting factors may become decisive. In addition, human activity is itself affected by environmental factors, and the two are often inseparable (although not indistinguishable) when analyzing their impact on species distribution.

The areas occupied by the northern and southern wolf nuclei are similar in some characteristics, namely low water availability, moderate to high environmental energy and high continentality (Table 2), which may be related to ecological factors. Continentality, for example, is related with winter severity, which can favour wolves by debilitating their natural prey (Mech et al. 1988*a*).

Although wolf distribution has been associated negatively with road density by several authors (e. g. Thiel, 1985; Mech *et al.*, 1988b; Mladenoff *et al.*, 1995), none of the three models presented here identify the distance to highways as positive for wolf presence. These results agree with Mech (1995), who noted that, in North America, wolf recovery after official protection also led to the colonization of much more open, accessible and densely populated areas with greater road density. Also Jędrzejewski *et al.* (2004) found no influence of roads on the distribution and abundance of wolves in Northern Poland, although there was a negative influence of highways and human settlements. The main menace posed by roads seems to come not from accidents, but rather from accessibility to humans and types of land use (see also Thiel 1985).

Despite some environmental similarities between the northern and southern wolf areas, our results suggest that the two nuclei show different spatial responses to the environmental constraints. There is a possibility that wolves from the northern nucleus have adapted to live in more humanized landscapes and to use artificial food sources, which may have helped their expansion. Several studies have indicated that, although wolves tend to avoid the proximity to densely populated areas, they start to tolerate them when their population grows or when good habitat becomes scarcer (Mech *et al.*, 1988*b*; Mech, 1995; Mladenoff *et al.*, 1995). Wolves have, in general, an extraordinary capacity to adapt and re-colonize areas if they are given the chance. Route and Aylsworth (1999) identified direct persecution and habitat destruction, caused by road building and the expansion of agricultural and urban areas, as the main threats to wolves at a global level, but they also pointed out that, given prey abundance and non excessive persecution, wolf populations recover rapidly.

The southern nucleus, in spite of its localized distribution, does not seem limited by space but rather by environmental factors related to energy availability, humidity and environmental disturbance. The model obtained for this nucleus points to a metapopulational structure in the spatial configuration of favorable areas, with the species occupying only two separated fractions of the main, core, and continuous favorable patch (see Figure 1c).

Connectivity seems to be of lower relevance for the northern wolf populations. The influence of spatial situation on the northern wolf nucleus reflects autocorrelation in distribution, which limits wolf presence to the vicinity of other presence areas. This autocorrelation in wolf distribution cannot be explained solely by autocorrelation in the analysed environmental predictors, since, in such case, purely spatial variables such as latitude and longitude would not add a significant predictive ability to the model. Consequently, this pure spatial effect may be interpreted as a consequence of aspects of this population's own dynamics, namely dispersal capacity (Legendre 1993, Barbosa et al. 2003). This may indicate a trend of the northern nucleus to keep expanding its range, as it is not limited only by environmental factors but also by the proximity of actual areas of presence.

Other top predators inhabiting the Iberian Peninsula, such as the Iberian lynx (*Lynx pardinus*) and the Iberian imperial eagle (*Aquila adalberti*), are in danger of extinction after restricting their diets to a staple prey, the wild rabbit (*Oryctolagus cuniculus*), whose populations have suffered strong declines (e. g. Real et al. 2009). In contrast, the ability of wolves to adapt their feeding habits has allowed them to recover notably after protective measures were taken. However, the models obtained here suggest that this process may not be generalizable to both Iberian wolf nuclei, as the southern nucleus is facing lack of connectivity between environmentally favorable areas. Given that species with disjunct distributions can show subspecific variation in their habitat requirements, due to local adaptation (Stockwell and Peterson, 2002; McPherson *et al.*, 2004), the plans for conservation and management of wolf populations in the Iberian Peninsula should take into account that the northern and southern nuclei have different biogeographical (and probably also ecological) traits, and thus should be treated independently.

A similar situation was recently observed for the two extant subspecies of Iberian wild goat *Capra pyrenaica* (Acevedo and Real 2011). Habitat fragmentation may be in the origin of disjunct distributions, when the movement between suitable fragments is disrupted. Species with disjunct distributions often show different responses to environmental gradients due to local adaptations (Stockwell and Peterson, 2002; McPherson et al., 2004), and they can become divergent enough for different genetic lineages to evolve. Such biogeographic drift might give rise to sub-speciation. Studies on spatial genetic structure of gray wolves have included the Iberian population, but with sampling limited to the northern nucleus (Pilot et al. 2010, vonHoldt et al. 2011). It would be interesting to genetically compare the wolves from the northern and the southern Iberian nuclei.

Disjunct distributions have relevant implications for species conservation and management, since they have different biogeographical, and probably ecological, characteristics, and should thus be considered independently when management strategies are designed (Acevedo and Real 2011). Disjunct distributions are also a crucial aspect to consider when transferring models in space or time, since under these circumstances the predictive capacity of the models may be strongly reduced under new scenarios (Barbosa et al. 2009, Acevedo and Real 2011).

ACKNOWLEDGMENTS

This study was originally included in A. M. B.'s PhD Thesis, presented in 2006 at the University of Málaga and financed by a doctoral grant from the Portuguese Fundação para a Ciência e a Tecnologia (SFRH/BD/4601/2001), co-financed by the European Social Fund. Further financial support was received from an FCT Post-Doctoral Fellowship (SFRH/BPD/40387/2007) and from Delta Cafés through the financing of the "Rui Nabeiro" Biodiversity Chair. Claudia Soares, Ana Filipa Alves, Francisco Álvares, and Rita Carreira / Grupo Lobo (Portugal) provided useful information on Portuguese wolf data. Yolanda Cortés, Francisco Álvares, and Juan Carlos Blanco made critical comments on earlier drafts of the manuscript.

REFERENCES

Acevedo and Real 2011 Biogeographical differences between the two *Capra pyrenaica* subspecies, *C. p. victoriae* and *C. p. hispanica*, inhabiting the Iberian Peninsula: Implications for conservation. *Ecological Modelling* 222: 814-823.

Araújo, M. B. and P. H. Williams, 2000. Selecting areas for species persistence using occurence data. *Biological Conservation* 96: 331-345.

Araújo, M. B., W. Thuiller, P. H. Williams and I. Reginster, 2005. Downscaling European species atlas distributions to a finer resolution: implications for conservation planning. *Global Ecology and Biogeography*, 14: 17-30.

Arberas, E., Blanco J. C., Campos M. A., Cortés Y., Gortazar C., Herrero J., Onrubia A., Palacios B., Palomero G. and Saenz De Buruaga M. 1999. Situación del lobo en España: una réplica. *Quercus*, 160: 62-64.

Arntzen J.W. and Espregueira Themudo, G. (2008) Environmental parameters that determine species geographical range limits as a matter of time and space. *Journal of Biogeography* 35: 1177–1186.

Austin, M. P., 2002. Spatial prediction of species distribution: an interface between ecological theory and statistical modelling. Ecological Modelling, 157: 101-118.

Barbosa, A. M., Real, R., Olivero, J. and Vargas, J. M. 2003. Otter (*Lutra lutra*) distribution modeling at two resolution scales suited to conservation planning in the Iberian Peninsula. *Biol. Conserv.* 114: 377-387.

Barbosa A.M., Real R. and Vargas J.M. (2010) Use of coarse-resolution models of species' distributions to guide local conservation inferences. *Conservation Biology* 24: 1378-1387.

Barbosa, A.M., Real R. and Vargas J.M (2011) Erratum to "Transferability of environmental favourability models in geographic space: The case of the Iberian desman (Galemys pyrenaicus) in Portugal and Spain" [Ecological Modelling 220 (2009) 747–754]. *Ecological Modelling* 222: 1067.

Barbosa A. M. and Real R. (2012) Applying fuzzy logic to comparative distribution modelling: a case study with two sympatric amphibians. *The Scientific World Journal*, Article ID 428206:10pp. (doi:10.1100/2012/428206).

Barbosa, A. M. and, R. Real and J. M. Vargas (2009) Transferability of environmental favourability models in geographic space: the case of the Iberian desman (Galemys pyrenaicus) in Portugal and Spain. *Ecological Modelling* 220:747-754.

Benjamini, Y. and Yekutieli, D. (2001) The control of the false discovery rate in multiple testing under dependency. *Annals of Statistics*, 29, 1165–1188.

Berger, J. 1999. Anthropogenic extinction of top carnivores and interspecific animal behavior; implications of the rapid decoupling of a web involving wolves, bears, moose and ravens. *Proceedings of the Royal Society of London* 266: 2261-2267.

Blanco, J. C. and Cortés, Y. (2002). *Ecología, censos, percepción y evolución del lobo en España: análisis de un conflicto.* Sociedad Española para la Conservación y el Estudio de los Mamíferos (SECEM). 176 pp.

Blanco, J. C., Cuesta, L., Reig, S. (1990). *El lobo (*Canis lupus*) en España. Situación, problemática y apuntes sobre su ecología.* Colección Técnica, ICONA. Madrid. 130 pp.

Blanco, J. C., Reig, S., Cuesta, L. (1992). Distribution, status and conservation problems of the wolf *Canis lupus* in Spain. *Biological Conservation,* 60: 73-80.

Boitani, L. 2000. *Action Plan for the conservation of the wolves (*Canis lupus*) in Europe.* Convention on the Conservation of European Wildlife and Natural Habitats. Council of Europe. Strasbourg. 86 pp.

Bulluck, L., E. Fleishman, C. Betrus y R. Blair, 2006. Spatial and temporal variations in species occurrence rate affect the accuracy of occurrence models. *Global Ecology and Biogeography*, 15: 27-38.

Cabral, M. J. (coordinator), J. Almeida, P. R. Almeida, T. Dellinger, N. Ferrand de Almeida, M. E. Oliveira, J. M. Palmeirim, A. L. Queiroz, L. Rogado y M. Santos-Reis (editores), 2005. *Livro Vermelho dos Vertebrados de Portugal*. Instituto de Conservação da Natureza, Lisboa.

Corsi, F., Duprè, E., Boitani, L., 1999. A large-scale model of wolf distribution in Italy for conservation planning. *Conservation Biology* 13, 150-159.

Ellegren, H., 1999. Inbreeding and relatedness in Scandinavian grey wolves Canis lupus. *Hereditas*, 130: 239-244.

Fico R., Morosetti G. and Giovannini A. 1993. The impact of predators on livestock in the Abruzzo region of Italy. *Rev. sci. tech. Off. int. Epiz.* 12: 39-50.

Fielding A. H. and Bell J. F. 1997. A review of methods for the assessment of prediction errors in conservation presence/absence models. *Environmental Conservation*, 24: 38-49.

Font, I., 1983. *Atlas climático de España*. Instituto Nacional de Meteorología, Madrid.

Font, I., 2000. *Climatología de España y Portugal*. Ediciones Universidad de Salamanca.

García, L. V. 2003. Controlling the false discovery rate in ecological research. *Trends in Ecology and Evolution* 18:553-554.

GRASS Development Team (2010) Geographic Resources Analysis Support System (GRASS) Software. Open Source Geospatial Foundation Project. Available from .

Guisan, A. and Zimmermann N. E., 2000. Predictive habitat distribution models in ecology. *Ecological Modelling*, 135: 147-186.

I. C. N., 1997. *Conservação do lobo em Portugal. Projecto realizado ao abrigo do programa Life. Relatório final.* Instituto da Conservação da Natureza, Lisboa. 231 pp.

I. G. M. E., 1979. *Mapa hidrogeológico Nacional. Explicación de los mapas de lluvia útil, de reconocimiento hidrogeológico y de síntesis de los sistemas acuíferos*. Segunda edición. Instituto Geológico y Minero de España, Madrid.

I. G. N., 1999. *Mapa de carreteras. Península Ibérica, Baleares y Canarias.* Instituto Geográfico Nacional / Ministerio de Fomento, Madrid.

IUCN (2010) *IUCN Red List of Threatened Species*. Version 2010.4. Available from http://www.iucnredlist.org/initiatives/mammals (downloaded in March 2011).

Jędrzejewski, W. Niedziałkowska, M., Nowak S. and Jędrzejewska B. 2004. Habitat variables associated with wolf (*Canis lupus*) distribution and abundance in northern Poland. Diversity and Distributions 10, 225-233.

Jiménez-Valverde A. and Lobo J.M. (2006) The ghost of unbalanced species distribution data in geographical model predictions. *Diversity and Distributions* 12: 521–524.

Legendre, P., 1993. Spatial autocorrelation: trouble or new paradigm? *Ecology* 74, 1659-1673.

Liu C., White M. and Newell G. (2009) Measuring the accuracy of species distribution models: a review. 18th World IMACS / MODSIM Congress, Cairns, Australia, 13–17 July 2009. Pp. 4241–4247.

Lobo, J.M., Jiménez-Valverde, A. and Real, R. (2008) AUC: a misleading measure of the performance of predictive distribution models. *Global Ecol. Biogeogr.* 17, 145-151.

McCullagh, P. and J. A. Nelder. 1989. *Generalized Linear Models*. Chapman and Hall: London.

McPherson, J., W. Jetz and D. J. Rogers, 2004. The effects of species' range sizes on the accuracy of distribution models: ecological phenomenon or statistical artefact? *Journal of Applied Ecology*, 41: 811-823.

Mech, D. L. 1995. The challenge and opportunity of recovering wolf populations. *Conservation Biology* 9(2): 270-278.

Mech, L. D., Fritts, S. H., Paul, W. J., 1988a. Relationship between winter severity and wolf depredations on domestic animals in Minesotta. *Wildl. Soc. Bull.* 16, 269-272.

Mech, L. D., Fritts, S. H., Radde, G. L., Paul, W. J., 1988b. Wolf distribution and road density in Minesotta. *Wildl. Soc. Bull.* 16, 85-87.

Mech, L. D. (1995). The challenge and opportunity of recovering wolf populations. *Conservation Biology*, 9(2): 270-278.

Mladenoff, D. J., T. A. Sickley, R. G. Haight, and A. P. Wydeven. 1995. A regional landscape analysis and prediction of favorable gray wolf habitat in the northern great lakes region. *Conservation Biology* 9:279-294.

Montero de Burgos, J. L. and J. L. González-Rebollar, 1974. *Diagramas bioclimáticos*. ICONA, Madrid. Pp. 45-48.

Muñoz-Cobo, J., Calvo, J. A., Carrasco, R. (1999). Bases para la elaboración del plan de conservación del lobo (Canis lupus L., 1758) en Andalucía.

Pp. 25-36. En: I*nvestigación y desarrollo medioambiental en Andalucía*. Junta de Andalucía y Univ. Jaén.

Palomo, L. J. and J. Gisbert (eds.), 2002. *Atlas de los mamíferos terrestres de España*. Dirección General de Conservación de la Naturaleza-SECEM-SECEMU, Madrid.

Pilot M. et al. (2010) Phylogeographic history of grey wolves in Europe. *BMC Evolutionary Biology*, 10: 104.

Quantum GIS Development Team (2010) Quantum GIS Geographic Information System. Open Source Geospatial Foundation Project. Available from

R Development Core Team (2009) R: a language and environment for statistical computing. R Foundation for Statistical Computing. Available from http://www.r-project.org (accessed January 2009).

Real R., Barbosa A.M. and Vargas J.M. (2006) Obtaining environmental favourability functions from logistic regression. Environmental and Ecological Statistics 13:237-245.

Real, R., Barbosa, A. M. and, A. Rodríguez, F. J. García, J. M. Vargas, L. J. Palomo and M. Delibes (2009) Conservation biogeography of ecologically-interacting species: the case of the Iberian lynx and the European rabbit. *Diversity and Distributions* 15:390–400.

Robertson, M. P., Peter C. I., Villet M. H. and Ripley B. S., 2003. Comparing models for predicting species' potential distributions: a case study using correlative and mechanistic predictive modelling techniques. *Ecological Modelling*, 164: 153-167.

Route, B. and Aylsworth, L. 1999. 1999 World Wolf Status Report. *International Wolf* 9 (3): Special Insert. International Wolf Center, Ely (Minnesota).

Sastre, P., Roca, P., Lobo, J.M., and EDIT co-workers (2009) A geoplatform for improving accessibility to environmental cartography. *Journal of Biogeography* 36: 568.

Stockwell, D. R. B. and Peterson A. T., 2002. Effects of sample size on accuracy of species distribution models. *Ecological Modelling*, 148: 1-13.

Thiel, R. R., 1985. Relationship between road densities and wolf habitat suitability in Wisconsin. *American Midland Naturalist*, 113: 404-407

Treves, A., Naughton-Treves, L., Harper, E. K., Mladenoff, D. J., Rose, R. A., Sickley, T. A. and A. P. Wydeven, 2004. Predicting human-carnivore conflict: a spatial model based on 25 years of wolf predation on livestock. *Conservation Biology* 18: 114-125.

U. S. Geological Survey, 1996. GTOPO30. *Land Processes Distributed Archive Center*, http://edcdaac.usgs.gov/gtopo30/gtopo30.asp.

Vilà, C., A.-K. Sundqvist, Ø. Flagstad, J. Seddon, S. Björnerfeldt, I. Kojola, A. Casulli, H. Sand, P. Wabakken and H. Ellegren. 2003. Rescue of a severely bottlenecked wolf (*Canis lupus*) population by a single immigrant. *Proc. R. Soc. Lond.* B 270: 91–97.

von Holdt, B. M., J. P. Pollinger, D. A. Earl, J. C. Knowles, A. R. Boyko, H. Parker, E. Geffen, M. Pilot, W. Jedrzejewski, B. Jedrzejewska, V. Sidorovich, C. Greco, E. Randi, M. Musiani, R. Kays, C. D. Bustamante, E. A. Ostrander, J. Novembre, and R. K. Wayne 2011. A genome-wide perspective on the evolutionary history of enigmatic wolf-like canids. *Genome Research*, 21, 1294-1305.

In: Carnivores
Editors: F. I. Álvares and G. E. Mata
ISBN 978-1-62257-323-3

Chapter 6

Factors Affecting Small and Middle-Sized Carnivore Occurrence and Abundance in Mediterranean Agricultural Landscapes: Case Studies in Southern Portugal

Filipe Carvalho, Ana Galantinho and António Mira
Unidade de Biologia da Conservação,
Universidade de Évora, Évora, Portugal
Grupo de Ecossistemas e Paisagens Mediterrânicas –
Instituto de Ciências Ambientais e Agrárias Mediterrânicas,
Universidade de Évora - Núcleo da Mitra, Évora, Portugal

Abstract

Carnivores have a key role in ecosystems and their populations are declining at an increasing rate. Habitat loss and degradation through agricultural practices are among the most serious menaces affecting carnivore survival. Agricultural landscapes are dominant in Western Europe and in several places worldwide are changing quickly into more intensive practices. Therefore, it is important to understand how small carnivore occurrence and abundance is affected by these changes in order to find ways to manage agricultural and grazing systems in a sustainable way, allowing both biodiversity and production to co-exist. We present three case studies in Natura 2000 areas, mainly covered by private agricultural land (livestock, cereal crops, oak, pine and eucalyptus plantations) in southern Portugal. In areas of dominant traditional agro–silvo–pastoral systems (montado) small and middle-sized carnivores tend to occur in a mosaic of montado and shrubs. We found that the occurrence of one of the forest species was positively related with the density of trees and shrubs, soil organic matter content, and Shannon's index of vegetation vertical diversity. The presence of livestock and extension of game-estate areas also seem to influence carnivore occurrence. In areas of extensive cereal crops, the presence of shrubs and age of forest plantations play an important positive role for the carnivore community. Mainly our results suggest that maintaining a sustainable mosaic embracing montado, shrubland and open land areas, may allow higher species richness and abundance by enhancing connectivity between crucial areas. The implementation of this kind of agricultural practices considering the landowners needs is the key issue to achieve the main carnivore conservation goals in Southern Portugal.

INTRODUCTION

Few areas of the Earth have not yet been affected by human activities (Sunquist and Sunquist, 2001). Worldwide, thousands of hectares of land have been converted to agriculture due to the increasing human population demands. One of the greatest challenges of the 21st century is food security since in 2008 nearly 15% of the world's population was considered undernourished and diminishing by 50% the poverty and hunger, by 2015, requires a greater investment in agriculture in developing countries (OECD-FAO, 2009). Currently, 30 countries around the world require external assistance for cereal supplies as a result of natural disasters, conflict or insecurity, and economic problems (GIEWS-FAO, 2009). About 4.3 billion of the 13.2 billion hectares of the world's land surface are moderately to highly suitable as rain-fed cropland and accounting for currently cultivated cropland (around 1.4 billion hectares) and for forested and urban/protected areas, 1560 million hectares are still available for cropland expansion (OECD-FAO, 2009).

In the past the agricultural expansion has been the main driving force of ecosystem change and since the middle of the 20th century the technological and economic incentives allowed for a quicker conversion and intensification of agriculture causing a pervasive decline in farmland biodiversity in recent decades (Benton et al., 2003; Groom and Vynne, 2006). Presently, the decline of biodiversity associated with agricultural practices is mainly due to the intensification of agriculture, the abandonment of marginally productive but high nature value farmland, and the changing scale of agricultural operations (Henle et al., 2008). Agricultural practices (including crop, livestock farming and timber plantations) are one of the three main causes of habitat fragmentation, degradation and loss (Groom and Vynne, 2006) which ultimately is the highest threat to terrestrial vertebrates and biological diversity (Crooks 2002; Michalski and Peres, 2005). Most of the agricultural practices destroy much of the original vegetation cover leading to changes in the community structure and composition, diversity and behaviour of the native fauna (Lyra-Jorge et al., 2008).

Nevertheless, the effect of agricultural practices on biodiversity depends upon the characteristics of the landscape and on the requirements of the existing species. Agricultural landscapes can be very different depending on soil, water, climate, slope and human management (Henle et al., 2008). On the other hand, a specialist species may become locally extinct due to the loss or fragmentation of a particular habitat, while a generalist species may benefit by the opportunities a new habitat may bring such as supplementary food and loss of specific competitors or predators (Crooks, 2002; Gehring and Swihart, 2003; Acosta-Jamett and Simonetti, 2004).

A focus on protecting places that simultaneously present high percentage of endemic species and greater areas of habitat destroyed is needed to reconcile biodiversity conservation and world development in the present context of increasing demands for agricultural expansion and intensification which endanger the conservation of biodiversity worldwide. In this sense the Mediterranean Basin is one of the 25 Global Biodiversity Hotspots (Myers *et al.*, 2000; Cuttelod et al., 2008). This region besides having an extremely high diversity of flora and fauna is also a mosaic of natural and cultural landscapes (Cuttelod et al., 2008). For thousands of years, human activities and natural ecosystems have coexisted in the Mediterranean in a complex way originating high environmental diversity and fragmentation (Maiorano et al., 2006). Currently most of the remaining ecosystems are highly managed and therefore conservation efforts must be directed to this mosaic of mostly hostile habitat, spatial and temporal heterogeneous created by agricultural activities (Macdonald and Rushton, 2003).

In an effective conservation process, human activities should be considered as part of the system since traditional agriculture and pasture allow for high levels of biodiversity (Maiorano et al., 2006). In the Mediterranean

Basin, agroforestry systems have shaped the landscape and the vegetation, being able to overcome the highly variable Mediterranean climate (Joffre et al., 1999).

One of the most representative habitats that contributed for the hotspot designation of the Mediterranean Basin is the open arable area of South Iberia, also known as pseudosteppes. This habitat found in Portugal and Spain comprises the major winter quarter for European migratory birds (Moreira et al., 2005). This high bird diversity depends on low intensive cereal croplands where a rich and abundant assemblage of insects is commonly found. In the last decades, many of these areas have been facing an agricultural intensification mainly in the most productive land, being the remaining areas abandoned, with the consequent shrub encroachment and afforestation with exotic species (Moreira et al., 2005; VanDoom and Bakker, 2007; Reino et al., 2009).

The major consequences of these management practices (or the lack of them) may be the decline of most of the winter migratory birds in Europe, once their resting area used during the winter migration is lost and predation pressure increases associated to the changes in land use (Shapira et al., 2008; Pita et al., 2009).

The montado (equivalent to the Spanish *dehesa*) is the dominant agro-silvo-pastoral system in South-western Iberian Peninsula and 33% of the worldwide cork oak range occurs in Portugal (Pinto-Correia, 1993). This traditional landscape is dominated by cork (*Quercus suber*) and holm (*Quercus rotundifolia*) oak trees that emerge from a grassland or shrubland matrix. In the montado although there is a repetition of similar elements, the different densities of trees and livestock, the rotational system of exploration and the different development of the vegetation allow for heterogeneity to exist in an extent necessary for maintaining a high biodiversity (Pinto-Correia and Vos, 2004; Rosalino et al., 2009). The combination between tree stands and grassland areas also enhances higher levels of plant and animal biodiversity (Ramirez and Díaz, 2008). This complex system of Mediterranean landscape remained multifunctional at several levels, nevertheless, in recent decades the montado system has gone through a process of intensification and consequent simplification of land use (Pinto-Correia and Vos, 2004).

The increase of tree diseases related with alterations in soil conditions and destructive management practices (e.g. understory removal), the abandonment of the traditional explorations and the reforestation using exotic species (e.g. *Eucalyptus* spp.) endangers the traditional balance between the montado components that previously assured the maintenance of high levels of biodiversity (Pinto-Correia and Vos, 2004; Santos-Reis et al., 2004).

Understanding and predicting the distribution, abundance and persistence of wild species in highly valuable habitats such as pseudosteppes and montado is a key element in conservation planning (Macdonald and Rushton, 2003).

Although mammalian carnivores have a historic resilience as a group, today many of these species are threatened with extinction (Gittleman et al., 2001). Even if several species are not considered presently threatened, most of the carnivores face a hard human persecution due to their known predatory behaviour, especially in agricultural areas where livestock production and game activities are a common practice (Ginsberg, 2001; Macdonald, 2004; Michalski and Peres, 2005).

Mammalian carnivores are wide-ranging and low density species and consequently are very sensitive to habitat loss and fragmentation (Oehler and Livaitis, 1996; Livaitis, 2001; Crooks, 2002; Riley et al., 2003; Lyra-Jorge et al., 2010). The activity patterns of carnivores are influenced by many factors, including habitat fragmentation, temperature variations, limitations of the visual system, risk of predation, competition, social and thermoregulatory behaviour (Zielinsky, 2000).

Carnivore' conservation must be a worldwide goal, due to their unique function in ecosystems as regulators of the trophic cascades (Gittleman et al., 2001; Mangas et al., 2008). Carnivores can be considered umbrella species due to their generally high home ranges that encompass the vital areas of several other species and therefore habitat management for carnivores may assure the conservation of high levels of biodiversity (Hilty et al., 2006).

Surveys regarding carnivores are time-consuming and expensive, which has discouraged many researchers to study these species, especially the big carnivores that have larger home ranges (Sunquist and Sunquist, 2001).

However, despite these logistical constraints, the ecological features of carnivores make them good models, if we aim to get a first glance on the effects of landscape fragmentation on wildlife. Indeed, they are excellent tools, if we want to evaluate the responses to fragmentation at different landscapes scales (local, regional and global) (Oehler and Livaitis, 1996).

On the other hand, small sized carnivores like the majority of the species occurring in Portugal due to their small home ranges and higher densities can be good surrogates to evaluate the effect of habitat changes on carnivore community, when compared with larger species. Some forest species as the American martens (*Martes americana*) have already been studied for these purposes with good and encouraging results in other regions (Hargis et al., 1999).

In this chapter, we will try to present and discuss the response of small and middle-sized carnivores to the two main types of agricultural activities that shape the landscape in southern Portugal: 1) the montado as an agro-silvo-pastoral ecosystem; and 2) the Mediterranean arable land as a mosaic of crops. These two landscapes are particularly important due to their extent, high biodiversity and high degree of threat by the trend in agricultural intensification accordingly to the European Union. Three case studies will be described in order to disentangle the major factors affecting carnivores' occurrence and abundance in these two important agricultural landscapes. These case-studies are particularly important because the data were gathered in Natura 2000 sites mainly composed by private property.

In these lands active agricultural management is directly related to local livelihoods in a rural socio-economic context dominated by extensive agricultural practices since many generations ago in Alentejo, southern Portugal. The first two case-studies are based on papers in preparation (Carvalho *et al*, *in prep*; Carvalho and Mira, *in prep*) and the third case study is based on a recently published paper (Galantinho and Mira, 2009).

CASE STUDIES

Forest Plantation Effects on Middle-Sized Carnivore Assemblages in an Open Arable Landscape: A Preliminary Approach

One of the most important agricultural landscapes prevailing in South Portugal is the open arable land, which is a result of extensive cereal croplands campaigns initiated in the last century (see Figure 1) (Moreira et al., 2005). During these campaigns, holm and cork woodlands were replaced by large open areas with cereal croplands. Consequently, several small and middle-sized carnivores considered habitat specialist as wild cats (*Felis silvestris*) and genets (*Genetta genetta*), which depend largely on woodlands and shrublands for resting and hunting sites, have been declining from those areas. On the other hand, generalist predators as feral dogs (*Canis familiaris*) and foxes (*Vulpes vulpes*) became predominant in these landscapes based on their capacity to explore all kind of habitats and food resources (Mangas et al., 2008; Pita et al., 2009).

In recent years, mainly after Portugal joined the European Union in 1986, due to the consequent new funding policies promoting production and agriculture intensification (e.g. cereal crops, livestock, and tree plantation) several changes have occurred in this landscape (Wade et al., 2008). In the last two decades, some of the economic incentives have lead to new forest plantations with umbrella pines (*Pinus pinea*) and holm (*Quercus rotundifolia*) and cork oaks (*Quercus suber*). Although tree plantations enhanced edge effects on otherwise open areas, these new plantations should be considered as causing soft edge effects when compared with the hard edge effects caused by old plantations already established in southern Portuguese landscapes (Reino et al., 2009).

The main goal of this study was to evaluate the effects of afforestation, since the early nineties, on the occurrence and abundance of small and middle-sized carnivore assemblages in the Special Protection Area (SPA) of Castro Verde. Secondly, we aimed to know which are the main landscape features influencing the occurrence of middle-sized carnivores in this kind of environment. The landscape is flat and dominated by an agricultural mosaic of cereal, fallow and cultivated fields originated by rotational dry cereal cultivation (Figure 1).

Until few years ago, tree cover was largely restricted to some old eucalyptus (*Eucalyptus* spp.) plantations and open holm oak (*Quercus rotundifolia*) woodlands grazed by livestock. The climate is typically Mediterranean, with hot and dry summers (average temperature of 24ºC) and cold winters (average temperature of 9ºC), the majority of the rainfall (500-600mm) occurring between October and March (Reino et al., 2009).

In 2005, we sampled carnivores through 'sign surveys' twice (summer and autumn) during one hour and half in each of the 60 sampling units located inside or in the surroundings of the study area (Figure 1). Fifty sampling units included edges between forested (of different ages and types) and open agricultural land, and ten consisted only in open agricultural areas (agricultural mosaic of cereal, fallow and ploughed fields). We chose a circular sampling unit of 300 m radius (~28 ha) to achieve a good balance between behavioural ecology of the species inhabiting the area, the size of the core areas of their known home ranges, the aims of the study and the available resources (Wilson and Delahay, 2001). Relative abundance indexes (RAI) for each species were estimated considering the total number of signs counted per each hour of search (Long et al., 2008). Scats of all species detected were identified based on their shape, size, location and smell.

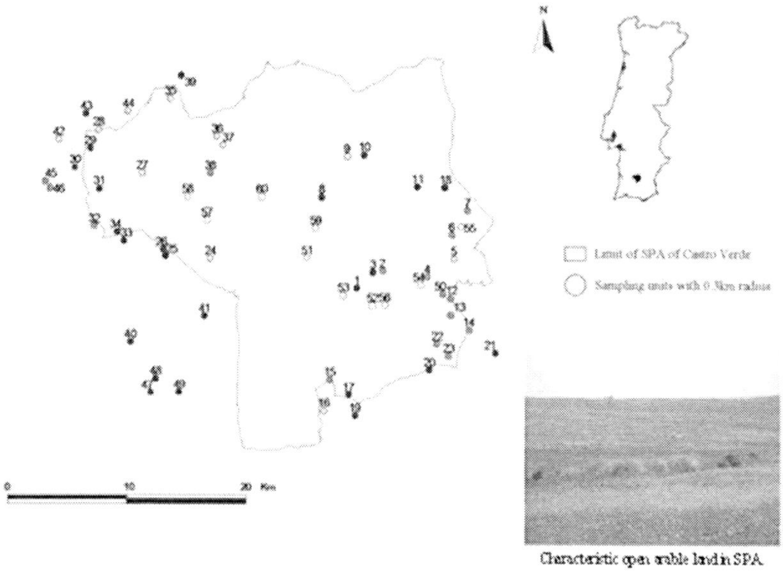

Figure 1. Study area in Castro Verde Special Protection Area. The location of the 60 sampling units and a detail image of the main habitat occurring in the area are also presented.

Any ambiguous identification of scats and footprints was discarded from data analysis (Barea-Azcón et al., 2007; Beja et al., 2009; Mangas et al., 2008). In each sampled plot, surveys were done along linear structures in the landscape such as dirt roads, fences, field edges and riverbanks, where signs of most small carnivore species are commonly found (Mangas et al., 2008). Other prominent structures in the landscape such as rocky areas and old trees were also surveyed to avoid false absences, because some species (e.g. genet) usually use them for latrine deposition (Espirito-Santo et al., 2007). On each sampling unit, several variables (Table 1) related to land use and landscape metrics (e.g. proportion of pastureland, urban areas and length of water streams) were computed using a GIS program *Arcview 3.2* (ESRI, 1999) and the *Patch analyst 2.2* extension (Eikie et al., 1999). Moreover, we analysed two categorical variables (both with four levels) concerning forest type and age of forest plantations. We compared the relative abundance and species richness of small and middle-sized carnivores between seasons using the Man-Whitney test. Furthermore, we compared those parameters among different types and ages of existing forests through analysis of variance (ANOVA) and/or nonparametric tests (Kruskal-Wallis test), whenever the Levene's test for homogeneity was violated (Sokal and Rohlf, 1997).

Table 1. Description and summary statistics of environmental variables (selected during exploratory analysis) used to examine carnivore responses to landscape structure and configuration in the study area

Acronym	Description	Mean ± S.D
DAMS	Proportion of area with dams (ha)	0.19 ± 0.07
DIRT_ROADS	Proportion of area with dirt roads (ha)	1.51 ± 0.11
ROADS	Proportion of area with roads (ha)	0.88 ± 0.35
RIP_AREAS	Proportion of area with riparian areas (ha)	0.79 ± 0.24
SHRUB	Proportion of area with shrubland (ha)	4.34 ± 1.39
REC_MT_SHRUB	Proportion of area with recent montado with shrubs (ha)	0.77 ± 0.62
OLIVE_GROVES	Proportion of area with olive groves (ha)	0.39 ± 0.17
PINE_FORESTS	Proportion of area with pine forests (ha)	9.06 ± 2.12
PASTURELAND	Proportion of area with pastureland (ha)	63.39 ± 2.41
URBAN_AREAS	Proportion of area with urban areas (ha)	0.15 ± 0.12
REC_FOREST	Proportion of area with recent forestation (ha)	17.47 ± 2.79
WATER_STR	Length of water streams per plot (m)	210.62 ± 13.90
FOREST_AGE	Age of forests inside each plot [a] (categorical)	2*
FOREST_TYPE	Type of forests inside each plot [b] (categorical)	2*
DIVERS	Shannon diversity index	1.21 ± 0.04

[a] Age of forests: 1 – without any type of forests (n =10); 2 – recent forests (< 5 years; n = 26); 3 – young forests (between 5 and 15 years; n = 11); 4 – Old forests (> 15 years; n = 13).

[b] Type of forests: 1 - without any type of forests (n= 10); 2 – oak forests (n = 24); 3 – eucalyptus forests (n = 11); 4 – pine forests (n = 15).

[*] For categorical variables the mode are presented.

Finally, we ran a multivariate redundancy analysis (RDA) for all species of carnivores on which reliable statistics analysis were possible (more than 10% of presences registered) (Zuur et al., 2007). All non collinear (R > 0.7) explanatory variables were considered in this analysis (Tabanick and Fidell, 2001).

Throughout the 180 hours of signs search during the summer and autumn of 2005, we observed 388 signs belonging to eight carnivore species: red fox, Eurasian badger (*Meles meles*), Eurasian otter (*Lutra lutra*), stone marten (*Martes foina*), Egyptian mongoose (*Herpestes ichneumon*), genet, domestic cat (*Felis catus*) and domestic dog. The carnivore assemblage was diverse and included almost all the common carnivore species that usually occur in southern Portugal (see e.g. Santos et al., 2008; Matos et al., 2009; Pita et al., 2009). The majority of the signs found were footprints (78.65%) followed by scats (19.01%) and less frequently others signs such as fur, skulls and prey remains (2.34%). The mean number of species registered per sampling area was 2.28 ± 1.21 (Mean ± Standard Deviation) of which 1.30 ± 1.06 were wild species and 0.98 ± 0.34 were domestic. The most common and abundant species were red foxes (RAI = 0.60 ± 0.01) and domestic dogs (RAI = 1.17 ± 0.17), matching respectively 28% and 54% of the signs recorded. Because of that only all predators, red foxes and feral dogs were used in comparison analysis. The remaining species occurred in less than 25% of the sampling sites, being the common genets (RAI = 0.02 ± 0.01) and feral cats (RAI = 0.02 ± 0.01) the least common.

No significant differences ($p > 0.05$) in relative abundance and species richness were found between seasons, neither for all predators and for the red fox and the feral dog considered individually (Man-Whitney test). However, the values were higher during the autumn season, perhaps because the rainy periods have lead to a mud accumulation in dirt roads, being the footprints easier to found and identify. Nevertheless, most of the juvenile dispersion, depending on carnivore species, starts on final summer and reaches the peak in autumn (Virgós et al., 2001; Matos et al., 2009), which may also contributed for these results for the wild species. Concerning the type and age of the forest plantation, we did not find significant differences for all predators' species, red foxes and feral dogs (Kruskal-Wallis test). The type of forest was expected to influence the diversity and the abundance of carnivores in the study area, because forests of exotic species are known to harbour less biodiversity than oak forests (Virgós et al. 2002; Pita et al. 2009). However, in our study carnivores were more abundant in areas of exotic species (pine and eucalyptus forests). This apparent paradox may be explained by the age of the forest plantations. In fact, most of the eucalyptus forests sampled were the oldest in study area, and presented highly developed undercover vegetation which harbours a great diversity of food resources and shelter sites, not found in the surrounding open agricultural matrix (Mangas et al., 2008). Conversely, the majority of the oak forests were planted less than five years ago. These recent oak plantations are structurally very similar to the open agricultural matrix and therefore the above result was not surprising. However, the more diverse assemblages of carnivores were found on the few old oak forests existing in study area (results not shown), which may reflect the tendency for a higher diversity of carnivores in mosaic landscapes (Gehring and Swihart, 2003; Matos et al., 2009; Pita et al., 2009). Clearly, the results highlight the fact that oldest forest plantations have the great carnivore relative abundances in the study area. Indeed, a positive linear relationship between age forest plantation and carnivore abundance was found (result not shown). This relationship was already documented in literature concerning the Mediterranean context (e.g. Pita et al., 2009).

The results of redundancy analysis (RDA) have shown that several other descriptors influence carnivore' community in SPA of Castro Verde: DIRT_ROADS, ROADS, FOREST_AGE, OLIVE_GROVES and URBAN AREAS (Figure 2).

The RDA showed a positive influence of the FOREST_AGE ($p < 0.05$) in the abundance of both red foxes and all wildlife predators taken together as previously discussed. On the other hand, feral dogs show only a weak positive correlation with it. In fact, dogs are known to be more dependent of human related activities as livestock herding and hunting (Pita et al., 2009). The abundance of dogs was also positively associated to OLIVE_GROVES which may reflect their location near small human settlements used mostly during the day in the study area. Nevertheless, wild carnivores are also positively related to OLIVE_GROVES probably because they may use the high availability of food resources (Revilla et al., 2000; Rosalino et al., 2005b) during the remaining hours.

The role of DIRT_ROADS as enhancers of landscape permeability for generalist carnivores, as feral dogs and red foxes, was also suggested by our results, especially for dogs. However, these results may reflect the easier way to find signs of carnivores on dirt roads, but even so this behaviour has already being described for carnivores and other mammal species (Gittleman et al., 2001; Brock and Kelt, 2004). Conversely, the other wild carnivores, as for instance genets, seem to avoid areas with high density of dirt roads and especially paved roads (ROADS) near bigger urban areas. Genets are forest specialists, depending on a structurally diverse vegetation (Galantinho and Mira, 2009; Carvalho and Mira, *in prep*). In the Mediterranean region, this species is known to avoid dirt roads crossing open lands probably due to the risk of being killed by feral dogs and red foxes (Palomares and Caro, 1999). Stone martens also have a similar behaviour, i.e. although they use dirt roads, these are avoided in open landscapes (Hargis et al., 1999; Virgós and García, 2002). Analysis also suggests that wild carnivores avoid urban areas as has been documented in other locations (e.g. Pita et al., 2009). Feral dogs seemed indifferent to the presence of large human settlements (URBAN_AREAS), although they have a well known commensally relationship with humans what may only be suggested by their relation with OLIVE_GROVES.

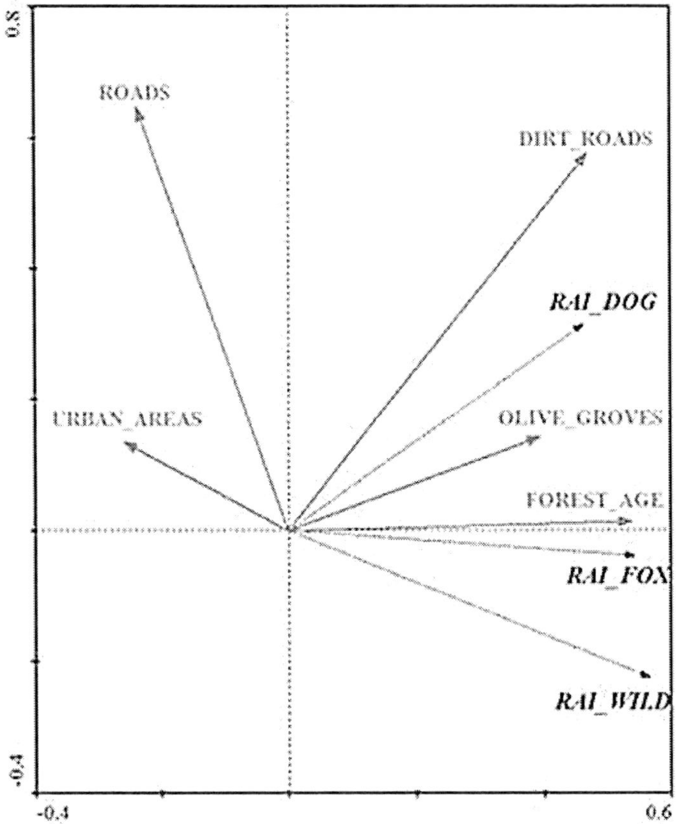

Figure 2. Ordination Biplot from Redundancy Analysis depicting the first two axes of the environmental variables relating to the relative abundance index (RAI) for the total wild carnivores species and the most common: dog and fox. Environmental variables (grey colour) are represented by solid lines and their acronyms (see Table 1). RAI for carnivore species (black colour) are represented by dashed arrows and their name.

The negative impact of ROADS may be a consequence of probable avoidance behaviour by some wild carnivores. Nevertheless, several carnivores are often seen along road verges searching and hunting for food, highlighting the risk of being road killed (Grilo et al., 2009).

Due to past deforestation for cereal crop purposes, and more recently by afforestation with exotic and native tree plantations, the specialist carnivores like the wild cat and the Iberian lynx (*Lynx pardinus*) and forest carnivores like the genet and stone marten have been decreasing or disappearing from South Iberia. The decrease in the abundance of the wild rabbit - a key prey species in the Mediterranean context - resulting from changes in land use practices, probably is another important cause for the decline of some specialist species. Although we did not present in this chapter the results obtained for lagomorphs surveys in the study area, rabbits were found only in scattered agricultural areas recently abandoned where shrub encroachment is in progress (Carvalho et al., *In prep*).

On the other hand, generalist species as red foxes and feral dogs have persisted through the years in southern Portugal and are now the most abundant carnivores in the study area. Our results agreed with other findings in the Mediterranean context, where forest plantations, especially in open arable landscapes seem to enhance abundance and diversity of wild generalist carnivores (Virgós, 2001; Virgós et al., 2002; Pita et al., 2009). In the future, this predictable increase in carnivore abundance and consequently in predator pressure could be a major conservation concern (Shapira et al., 2008) in SPA of Castro Verde, on which conservation value derives mostly from the important populations of steppe birds occurring there. In fact, one of the greatest causes of the decline of these species all over the world is ground nesting predation (Yanes et al., 1995; Disney et al., 2008).

Factors Influencing Carnivores' Distribution on a Montado Landscape

In the Iberian Peninsula, the montado landscape remains at a highly favourable conservation status, enabling the persistence of important populations of wild carnivores (Myers et al., 2000). The high diversity of the montado structure concerning the tree density and age, the mosaic of the understory vegetation alternating different types of shrubs with pastures and cereal crops promotes carnivore diversity and abundance. Particularly important for some species is the presence of an understory shrubland in some parts of the montado which provides shelter, food and resting sites. We should expect a higher richness and abundance of carnivores on this landscape when compared with surrounding extensive grassland or exotic reforestation areas (Conner et al., 1983; Virgós et al., 2002; Ramirez and Díaz, 2008; Mangas et al., 2008; Galantinho and Mira, 2009; Pita et al., 2009). The knowledge of habitat use, particularly the resting sites of a species is paramount for their conservation (Zabala et al., 2003). Obtaining an accurate estimate of the distribution and abundance of carnivores is very difficult and usually costly. Many biologists have to use binary data (presence/absence), based on detection rates obtained from less complex field studies to get a first glance on carnivores populations (Long et al. 2008).

The main aim of our research was to model the distribution of several sympatric middle-size carnivores on a montado landscape in Southern Portugal based on presence/absence data.

The study was conducted in Monfurado Natura 2000 Site (23,878 ha; 38º33´N, 8º 09´W) (Figure 3). The main threats to wildlife and conservation in Monfurado are habitat degradation through intensive livestock grazing, exotic reforestation, land abandonment, fire, and road construction (ICN, 2005). Monfurado only includes private properties and the montado corresponds to about 70 % of all the land uses.

Human presence is low and only 0.35% of the area is occupied by farmhouses and small villages. Game estates cover 62.3% of the area and on those areas predator control is legal and a common practice. The northern limit of the study area is bordered by a high-traffic national road (N114) and a recently constructed highway (A6). Numerous secondary paved and dirt roads cross the site.

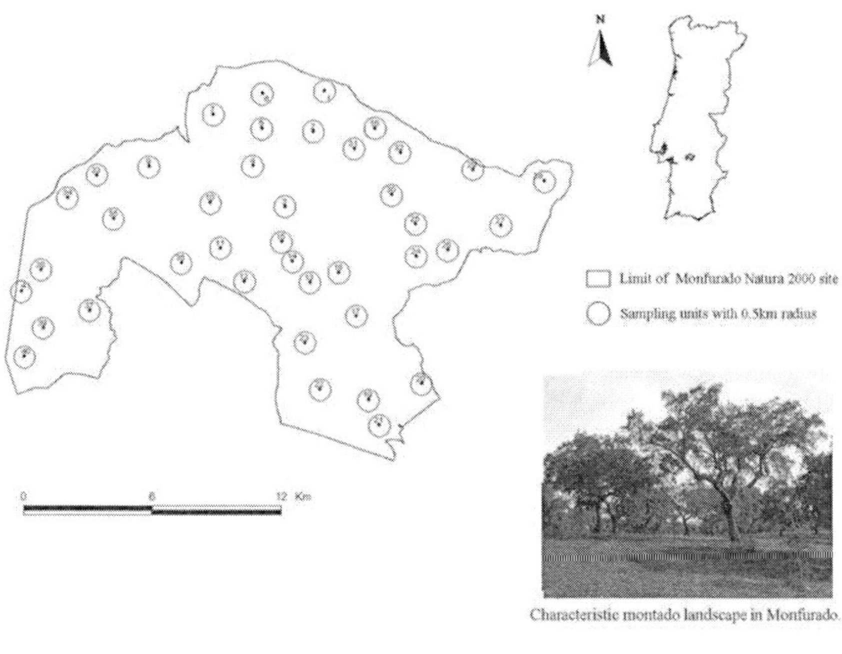

Figure 3. Study area (Site of Monfurado) location. The distribution of the 40 sampling sites and a detail image of the main habitat occurring in the area are also presented.

Monfurado has a typical Mediterranean weather with a dry and hot season (Rivas-Martinez and Loidi, 1999) from June to October, where monthly average temperatures range from 20 to 23°C. In the wet season, monthly mean rainfall is above 70 mm and mean temperatures range from 10 to 15°C. Altitude lies between 150 and 420 m.

The presence of middle-sized carnivores was accessed by scent stations (Diefenbach et al., 1994; Sargeant et al., 2003; Gompper et al., 2006; Long et al. 2008), being the data collected from October 2003 to February 2004. We chose 40 sampling sites covering all the study area (Figure 3).

Each sampling site consisted in a 0.5-km-radius buffer where five scent stations baited with oil sardines were set at least 200 meters apart from each other. This number of scent stations guarantees a high probability of recording a middle-sized carnivore species if present in each of the sampling sites.

Scent stations were implemented in fall and winter when low food availability occurs in southern Iberian Peninsula, which potentiates the carnivore response to bait (Barea-Azcón et al., 2007). This period corresponds also to the mating season of most carnivore species in the study area increasing the probability of attraction by scent bait (Macdonald and Barret, 1993). The land cover classification was based on the methodology described by VanDoorn and Pinto-Correia (2007) on VISTA research project (EVK2-2001-000365). This methodology, which considered several degrees of montado and shrubland densities, was applied to the land cover classification obtained from unpublished data collected in 2004 together with field confirmation and correction of land use changes (ERENA, 2004). The changes in the existing classification were made using a minimum mapping unit (m.m.u.) of 0.25 ha, which intend to reflect the resolution of how small carnivore movements may respond to habitat heterogeneity (Mazerolle and Villard, 1999). We used a GIS program *Arcview 3.2* (ESRI, 1999) and the *Patch analyst 2.2* (Eikie et al., 1999) extension to obtain the landscape metrics on each sampling unit. We considered, for instance, edge density (ED), perimeter of roads (ROAD_PER) and riparian habitats (RIP_PER).

ERENA also provided other independent variables like soil type, density and time spent by livestock in the areas, percentage of dead trees (MORTMONT) and diseases as charcoal canker (PCHAR) which promotes cork oak decline and death (see Table 2 for details).

Prior to analysis, the spatial autocorrelation was tested for each carnivore species between sampling sites, using the Moran's I and accounted for its significance with a z-test computed with the script of Lee and Wong (2001) for *Arcview* (ESRI, 1999).

Table 2. Variables groups, abbreviations, description and summary statistics for explanatory variables selected during exploratory analysis and included in models for predicting carnivore occurrence in Monfurado

Acronym	Description	Mean (± S.D)[c]
Landscape metrics		
AWMSI	Area weighted mean shape index (ha)	1.91 ± 0.48
ED	Edge density (km/km^2)	0.02 ± 0.01
NUMPP	Number of Patches	11.18 ± 4.34
ROAD_PER	Perimeter of roads per each sampling site (m)	7052.79 ± 4572.05
RIP_PER	Perimeter of riparian habitat per each sampling site (m)	1323.08 ± 1820.32
DIVERS	Index of plant structural diversity	0.21 ± 0.09
Land cover		
MAT1	Matorral with a low s.c[a] < 30 % (ha)	22.27 ± 26.49
MAT2	Matorral with a middle s.c 31-50 % (ha)	2.27 ± 6.83
MAT3	Matorral with a higher s.c > 51 % (ha)	4.31 ± 8.63
GRASS	Area with arable or grassland (ha)	46.55 ± 28.17
MONT1	Montado with a low c.c[b] < 30 % (ha)	17.64 ± 19.49
MONT2	Montado with a middle c.c 31-50% (ha)	22.20 ± 22.07
MONT3	Montado with a higher c.c > 51 % (ha)	27.85 ± 28.22
Oli	Olive groves (ha)	5.84 ± 13.92
MORTMONT	Percentage of dead trees in montado landscape (%)	0.29 ± 0.42
PCHAR	Trees affected by charcoal canker (%)	8.29 ± 5.74
Soil type		
S_ARG	Low unsaturated clayed soil (ha)	47.87 ± 28.13
S_LIT	Dry soils (ha)	21.96 ± 25.31
S_IC	Incipient soils (ha)	1.67 ± 3.44

[a] s.c = shrub cover; [b] c.c = crown cover; [c] S.D – standard deviation.

We employed a Binary GLM (generalized linear models) analysis to predict the presence/absence of middle-sized carnivores using landscape metrics, land cover and soil type as explanatory variables (Table 2) (Hosmer and Lemeshow, 2000; Long et al. 2008). The capacity of the independent variables to discriminate between the presence and absence of middle-sized carnivores was examined by univariate regression during the exploratory data analysis phase (Wald test; $p < 0.25$). Collinearity was tested between pairs of variables and only the variable biologically more valuable was retained for further analysis from highly correlated pairs ($r > 0.70$) (Tabachnik and Fidell, 2001). All the possible biological plausible two–way interactions, between significant or nearly-significant ($p < 0.1$) independent descriptors, were also considered (Hosmer and Lemeshow, 2000). During the Binary GLM models construction, variables selection was performed by the forward elimination procedure (Hosmer and Lemeshow, 2000). Evaluation of models fitting was based on the classification tables (proportion of cases correctly classified) and ROC curves (Manel et al., 2001).

Statistical modelling was done only for the species recorded in more than 25% of the sampling sites (Table 3). The Moran I results show that there were no significant spatial autocorrelation for any of the species ($-1< Z <1$; $p>0.05$) (Lee and Wong 2001).

The best Binary GLM models for all species correctly classified more than 75 % of cases and presented a useful adjustment by the ROC curve (useful accuracy = 0.7-0.9), except for the stone marten model which only showed a near acceptable accuracy (AUC stone marten = 0.67). Cross-validation validated all models (< 50 % error classification for $p < 0.05$), meaning that they can be used in other areas with similar landscapes.

The red fox occurred more often in fragmented areas [significant positive association (+ β) with AWMSI] (Eikie et al., 1999). Foxes are able to survive in open landscapes, so they are less susceptible to forest loss than other small and middle-size carnivore species (Virgós et al., 2002). Moreover, larger and more vagile species, like the fox, are more prepared to cross and hunt in deforested areas (Gehring and Swihart, 2003). However, even for these species the presence of shrubs and trees are very important for shelter and den sites (Virgós et al., 2002). In fact, areas with 31% to 50% of tree canopy cover (MONT2) have a significant positive association with the presence of red fox in Monfurado. Nevertheless, the interaction between MONT2 and AWMSI had a negative influence in the presence of the red fox, probably meaning that the effect of the landscape fragmentation on fox occurrence depends on the montado patches area. This may reflect a possible threshold in the fragmentation degree, even for a generalist species (Pita et al., 2009).

The presence of the genet was significantly associated to MAT1 (areas with shrub cover under 30%, see details on Table 2). Lower to moderate density values of shrubland are preferred by genets as shelter, resting and reproduction sites (Palomares and Delibes, 1988 and 1994; Virgós and Casanovas, 1997; Carvalho and Gomes, 2004; Galantinho and Mira, 2009). This habitat seems to have a great abundance of small mammals like rodents, and passerines that together represent the greatest amount of biomass in the diet of the genet in the Iberian Peninsula (Virgós et al., 1999; Rosalino and Santos-Reis, 2002; Carvalho and Gomes, 2004).

Table 3. Summary statistics of the final best models selected to predict the presence of middle-sized carnivore through scent stations in *Monfurado* study area

Red fox Model	β [a]	SE [b]	Wald [c]	Sig. [d]
AWMSI*MONT2	-0.119	0.062	3.708	0.054
MONT2	0.256	0.120	4.550	0.033
AWMSI	4.844	1.966	6.070	0.014
Intercept	-9.731	3.376	6.641	0.010
Genet Model				
DIVERS*PCHAR	1.026	0.408	6.314	0.012
MAT1	0.054	0.018	9.227	0.002
Intercept	-3.045	1.013	9.045	0.003
Stone marten Model				
MONT1*NUMP	0.003	0.002	3.729	0.053
AWMSI*MONT2	0.019	0.009	3.959	0.047
Intercept	-2.400	3.376	9.970	0.002
Badger Model				
PCHAR*S_ARG	-0.011	0.005	5.606	0.018
MORTMONT*OLI	-5.596	2.573	4.730	0.030
MORTMONT	5.162	2.635	3.839	0.050
Intercept	2.154	1.332	2.618	0.106

[a] β – Estimate regression coefficient; [b] S.E – Standard error; [c] Wald statistic; [d] Significance level.

Areas with a great diversity of vegetation and percentage of trees affected by charcoal canker are significantly associated with the presence of the genet. The index of plant structural diversity reflects the vertical structure of vegetation in the landscape, which is a crucial aspect for a forest carnivore as the genet. In fact, this carnivore has territorial habits, where faecal marking behaviour occurs normally in trees (cork and holm oaks), but also on the ground inside the core areas in a moderate density of shrubland (Palomares and Delibes, 1994; Virgós, 2001; Carvalho and Gomes, 2004; Galantinho and Mira, 2009). In Monfurado throughout the field work, we also found genet latrines in rocks, when no trees were found (Carvalho and Mira, *in prep*). The charcoal canker disease is derived by the contamination of a fungus denominated *Nummularia regia* (syn. *Biscogniauxia mediterranea*) and affects cork oak trees (Nugent et al., 2005). This fungus destroys the cork tree tissues (phloem and xylem), so the tree becomes dark (carbon colour) and hollow. Genets use those holes as shelter to rest, for reproduction and also for minimizing the risk of predation by dogs and red foxes (Palomares and

Delibes, 1994; Santos-Reis et al., 2004). The presence of the stone marten was positively associated with two interactions: MONT1 (tree canopy cover below 30%) and NUMP (number of patches); MONT2 and AWMSI. Both interactions deal with predictors directly related with forest cover and habitat fragmentation. As for the red fox, the stone marten model seems to reveal the existence of a threshold in the rate between matrix (open land) and forests above which the species disappear, as already stated for other carnivores (Oehler and Livatis, 1996; Hargis et al., 1999; Hilty and Merenlender, 2004; Pita et al., 2009). Stone martens, despiste their forest preferences, can tolerate human presence, whenever the associated extra food resources exist, like fig trees, apple trees and wild roses (Hargis et al. 1999; Rondinini and Boitani, 2002; Santos-Reis et al., 2004). Moreover, stone marten seems to prefer areas where small forest patches are embedded in open areas. These areas may be favourable to stone martens, because an increase in the abundance and diversity of small mammals and passerines associated with clear cuts occurs and still permits movements between patches inside the matrix (Virgós, 2001; Rondinini and Boitani, 2002; Virgós and García, 2002; Virgós et al., 2002). The badger was positively associated to MORTMONT (the percentage of trees affected by diseases in montado landscape). This species digs near tree roots to find food (e.g. insects) which is an easier task around sick or dead trees (Moore et al. 1999). Furthermore, the places surrounding dead trees are areas with more organic matter which enhances the abundance of the soil fauna (Revilla et al., 2000; Rosalino et al., 2003, 2005b). However, when some tree diseases affect orchards and olive groves, their fruit production might be compromised. Fruits are a very important food resource for generalist carnivores like the badger. Actually, our data showed that the probability of occurrence of badger decreases in areas where the percentage of diseases increases associated with olive grooves. The decrease in fruit availability especially in some seasons as in winter (e.g. olives) can be crucial for the dispersion of young badgers (Revilla et al. 2000; Rosalino et al., 2005b). In the Mediterranean region, the densities of badger populations are low, when compared with the populations from North Europe (e.g. UK) probably due to the high xericity in southern Iberian Peninsula, where factors influencing location of den sites must be critical. The negative interaction between charcoal canker (PCHAR) and loam soils (S_ARG) may reflect a higher degree of human disturbance. In fact, the charcoal canker affects highly trees from genera *Quercus* sp., which result in an economical constraint in study area concerning cork production. The higher human presence due to this disease control may affect den site fidelity by badgers, even if the density of den sites should be higher due to a good soil for digging dens (Mickvicius, 2002; Rosalino *et al.*, 2005a). However, badger may have to travel greater distances to find food safely (Revilla et al., 2001). Although the models obtained had high predictive power, we should interpret our results with caution and consider them as a tool which may state out trends about the way Iberian carnivores can be distributed in the montado landscape (Hilty et al., 2006). However, we think that these results are in concordance with others surveys realized in Iberian Peninsula for small and middle-sized carnivores (Virgós, 2001; Virgós et al., 2002; Mangas et al., 2008; Matos et al., 2009; Pita et al., 2009; Carvalho and Mira, *in prep*).

One of the main results highlighted, is that the presence of large montado landscape areas, together with a moderate shrubland density interspersed with open lands is one of the best habitats for most of the Iberian carnivore species (Virgós, 2001; Pita et al., 2009).

Montado landscapes depend on extensive human related activities and some of our results support the fact that human intervention is critical to maintain favourable conditions for carnivore occurrence (Ramirez and Díaz, 2008). Once identified the most significant factors influencing the presence of carnivores, this information should be incorporated in management plans aiming biodiversity conservation. We strengthen the fact that effective management actions in human-dominated landscapes should take place not only on the Natura 2000 sites but also outside these areas.

Human, Ecological and Livestock Influences on Genet (Genetta Genetta) Distribution on Mediterranean Farmland

The genet is widely distributed and common in Portugal (Santos-Reis and Mathias, 1998), and classified as ''Least Concern'' in Europe (IUCN, 2007). Evaluating the relative importance of specific issues of local management on the distribution of wild species is important if we wish to integrate farmland activities in species conservation plans. An abundant and common forest species like the genet may be a representative of the small and middle size carnivore community in montado dominated landscapes and may be used as surrogate to model the influence of ecological and human related activities on this community as a whole. Although each carnivore species have specific characteristics and needs they share resources (e.g. food, space) where they co-exist.

The presence/absence pattern of the genet in southern Portugal was evaluated considering natural landscape features and two important axes of rural management: livestock production and direct human influence.

We aimed to identify main factors or groups of factors influencing the distribution of the genet and especially how management practices influence its presence on Mediterranean farmland. As in the previous case-study the fieldwork was conducted in Monfurado a Site included in the Mediterranean biogeographic region of Natura 2000 Network in southern Portugal (Figure 3). Thirty six homogeneous habitat patches were used as sampling sites. These sites were selected accounting for habitat heterogeneity (agro–silvo-pastoral use, shrub structure, tree density) and representativeness in the study area. On three occasions (summer, autumn/winter and spring), from June 2003 to June 2004, every patch was surveyed walking on a path of approximately 1,000 m long and 25 m wide searching for signs (faeces and tracks) of genet. Latrines were searched in old trees (Palomares, 1993), fallen tree trunks (Queirós, 1989), old decaying roofs and walls (Gomes and Giraudoux, 1992), large rocks (Roeder, 1980; Palomares and Delibes, 1994; Virgós and Casanovas, 1997), and raised ground (Palomares, 1993; Virgós and Casanovas, 1997). Twenty-eight explanatory variables were collected or derived from a 500 m radius sampling circular plot centred in each of the patches (see for more details Galantinho and Mira, 2009).

The variables were grouped into three sets of descriptors: ecological, livestock, and human influence. Both livestock and human influence groups reflect different types of suspected perturbation. The ecological group included each of the most representative habitats (e.g. HMONT_HSHUB; MEADOWS), Shannon's diversity index (SDI), topographic variables (e.g. ITR; ELEV_RANGE; FLAT), soil organic mater (SOIL_OM) and trees affected by diseases (e.g. PCHAR). The livestock variables comprised density and time spent in the plots by sheep and cattle (e.g. SHEEP_D; SHEEP _T). The direct human influence accounted for distance to roads (DIST_ROAD) and villages (DIST_VILL), number of buildings present in the sampled areas (N_BUILD), area covered by small-game estates (SGAME_EST) and shrub cutting (SHRUB_CUT). Moran's I was used to test for autocorrelation in the genet data. The influence of each set of explanatory variables on the presence of the genet was assessed by the variation partitioning procedure proposed by Borcard et al. (1992), adapted to logistic regression models and extended to the three sets of variables. The best models for each set were chosen by information–theoretical model comparison (ITMC) (Burnham and Anderson, 2002).

Table 4. Variables groups, abbreviations, description and summary statistics for significant explanatory variables selected during exploratory analysis and included in models for predicting occurrence of the genet in Monfurado

Acronym	Description	(Mean ± SD) (0)	(Mean ± SD) (1)
Ecological			
HMONT_HSHRUB	High Montado canopy and shrub cover (both > 30%)	1.78 ± 0.88	1.33 ± 0.59
MEADOW	Meadows (%)	1.78 ± 0.73	1.50 ± 0.62
WATER_C	Watercourses (m/m2)	$3.3*10^{-4} \pm 3.8*10^{-4}$	$1.5*10^{-4} \pm 3.2*10^{-4}$
SDI	Shannon's Diversity Index	0.18 ± 0.09	0.25±0.08
PCHAR	Trees affected by charcoal canker (%)	8.22 ± 5.66	11.38 ± 6.24
SOIL_OM	Soil organic matter (%)	3.12 ± 1.25	4.17 ± 1.31
ITR	Index of the Topographic Roughness (surface area/flat area)	1.01 ± 0.01	1.01 ± 0.01
ELEV_RANGE	Elevation range (m)	56.07 ± 30.38	75.86 ± 34.04
FLAT	Flat Area (%)	0.08 ± 0.08	0.04±0.05
CARNIVORE	Carnivore abundance (n signs/m)	$1.17*10^{-3} \pm 1.11*10^{-3}$	$1.65*10^{-3} \pm 9.9*10^{-4}$
Livestock			
SHEEP_D1	Sheep density (0 n/ha)	6	3
SHEEP _D2	Sheep density (0-1 n/ha)	7	7
SHEEP _D3	Sheep density (>1 n/ha)	5	8
SHEEP _T1	Time spent by sheep in the field (0-1 months/year)	7	6
SHEEP _T2	Time spent by sheep in the field (1-8 months/year)	3	9
SHEEP _T3	Time spent by sheep in the field (>8 months/year)	8	3
Human influence			
SGAME_EST	Small-game estate (%)	0.70 ± 0.53	0.43 ± 0.57
DIST_ROAD	Minimum distance to roads (m)	1162.28 ± 759.18	1471.53 ± 832.41
N_BUILD_1	Buildings (n)	6	8
N_BUILD_2	Buildings (n)	6	8
N_BUILD_3	Buildings (n)	6	2

We present the mean ± standard deviation (Mean ± SD) of significant continuous variables and the number of patches per class of categorical significant variables with (1) (n = 18) and without (0) (n = 18) genet signs. The number of buildings is analysed as a categorical variable because it has only three possible values (Tabachnick and Fidell, 2001).

Regression coefficients were considered significant at the $P < 0.15$ level (Rawlings, 1988) and slightly higher P values were also used to account for variables of particularly important biological meaning (Hosmer and Lemeshow, 2000). From all the explanatory variables collected, fifteen were used to complete the procedure (Table 4).

Genet's presence/absence data was not significantly autocorrelated. Genets were present in half the patches surveyed and its occurrence seemed to depend more upon the ecological factors (pure effect: 30.3%) than on grazing (pure effect: 8.5%) and direct human influence (pure effect: 4.8%) (total variance captured: 56.5%). High cover and the complexity of the vegetation strata, and organic content rich soils seem to be the most important ecological factors influencing the occurrence of the species (Table 5). These factors are often associated to dense shrub and tree cover (Moreno et al., 2007; Moreno, 2008). High understory cover and old trees are important for the placement of latrines in cork oak woodland (Espírito-Santo et al., 2007) and as resting places during the day (Palomares and Delibes, 1988). A dense shrub cover often offers to genets, a higher abundance of the wood mouse (*Apodemus sylvaticus*) which is considered their main prey, besides shelter against predators and competitors (Cugnasse and Riols, 1984; Palomares and Delibes, 1988, 1994; Virgós and Casanovas, 1997; Virgós et al., 2001). Although cork oak trees are one of the most important sources of profit for landowners, shrub areas are considered as potentially dangerous as highly inflammable and as obstacles in the fire extinguishing process.

The livestock models suggested that moderate grazing intensity by sheep favours the presence of the genet (Table 5). The vegetation structure and species composition of plants and arthropods in shrub and grass layers depends on sheep grazing intensity (Marrs et al., 2007; Woodcock et al., 2005). Low to moderate levels of livestock grazing increase arthropod diversity (see Debano, 2006). Arthropods (e.g. coleoptera, orthoptera, and arachnidae) are the second most frequently consumed prey by the genet after mammals, in Europe (Virgós et al., 1999; Rosalino and Santos-Reis, 2002). Small mammals, birds, and dung beetles are important food items for the genet and are highly available in montados associated with the presence of livestock (Virgós and Casanovas, 1997).

Although moderate grazing may have a positive effect in an important food item, higher intensities of livestock have extremely negative effects in biodiversity (Fleischner, 1994; Dobkin et al., 1998; Krueper et al., 2003; Wheeler, 2008). Moreover, intensive stocking rates endanger tree renewal which is essential to the sustainability of the montado system once the sprouting young trees are consumed or destroyed.

However, grazing prevents land abandonment and is an important source of income associated to the montado system in the study area. Therefore stocking rates compatible with conservation values must be defined for this and other montado areas. Rotational livestock grazing is a traditional management option recommended to provide habitat for those species requiring short swards and intense grazing, as well as for large, diverse invertebrate assemblages in taller, older swards (Dennis, 2003).

The best human influence model suggested that the occurrence of the genet is lower in hunting estates (Table 5).

Hunting estates are profitable associations that rely greatly on the contributions of its members and occasional hunters and most certainly on the numbers of game hunted. The main goal of hunting management is to maintain a profitable number of game species and individuals. Often, management goals in these areas include lowering predator numbers. In Portugal, legal predator control consists in trapping and killing foxes and Egyptian mongooses (*Herpestes ichneumon*). However, predator control actions can influence and threaten non-target species, like the genet, once they are inherently not selective and often raise negative effects on the overall carnivore guild composition and diversity (Duarte and Vargas, 2001; Virgós and Travaini, 2005). In fact, cage-traps have to be inspected daily to release non-target species otherwise they will die due to lack of food and water. Moreover, often people believe that all small and medium size carnivores are an eminent threat to game-species and livestock and therefore illegally kill all trapped animals.

Other predator control actions are used in many estates although they have already been considered illegal for many years. Poison is one of the most dangerous techniques, because it affects many species of mammal, reptiles and birds. Therefore, environmental awareness campaigns directed to landowners, game-estate managers and hunters about the role of carnivores in the ecosystem should be a generalized measure.

The legal process of attributing credentials for using traps should require more strict rules for their placement, checking frequency, and, a better and more effective surveillance system (Duarte and Vargas, 2001; Virgós and Travaini, 2005).

Globally considering the best models for each set of variables and the variation partition procedure, genet's presence was highly associated with management strategies, and indicators of the ecosystem's well-being (for more details on all the models compared see Galantinho and Mira, 2009).

Table 5. Best partial models for the ecological, livestock and human influence sets

Explanatory Variables	B	SE	Sig.
Ecological			
HMont_HShrub	2.852	1.954	0.144
SDI	8.773	5.442	0.107
Soil_OM	0.749	0.358	0.027
Constant	-5.251	2.069	0.011
Livestock			
Sheep_T			0.087
Sheep_T2	1.253	0.868	0.149
Sheep_T3	-0.827	0.876	0.345
Constant	-0.154	0.556	0.782
Human influence			
SGame_Est	-0.901	0.636	0.157
Constant	0.503	0.490	0.304

Best partial models, selected based on ITMC are reported along with the effect sizes, their precision and associated *P*-values. The interpretation of the regression coefficient for categorical variables is based on the first category (indicator).

Soil organic matter content is the most significant predictor in both partial and full models. The maintenance of soil organic matter has been an important goal of several research activities and programmes concerning environment conservation to assure the sustainability and productivity of the natural and agricultural systems (Wolters, 2000).

Certainly relevant variables were not analysed since 43.5% of the variance in the presence/absence pattern of the genet was not explained by the three sets of variables used. Probably, including prey availability and microhabitat characteristics would improve the explained variance of our models once predators have to assess and choose patches for prey availability (Stephens and Krebs, 1986) and microhabitat is known to be significant for the placement of latrines in the montado landscape (Espírito-Santo et al., 2007). Nonetheless, we explore issues regarding local farming management and soil conservation practices that are important for the sustainability of agro-silvo-pastoral activities and wildlife management in Mediterranean environments.

This is especially relevant for landowners to understand and engage voluntarily in conservation strategies that can initially be considered by them economically disadvantageous (Jacobson et al., 2003) such as maintaining shrub areas and diminishing grazing intensity. However, developing strategies to involve landowners in conservation actions should and can go beyond economic incentives once many of them are emotional attached to the land they own. Moreover, pure economic incentives (e.g. subsidies, tax reduction) do not have a long-term existence and effectiveness while the effects of social commitment, recognition and problem avoidance can engage emotionally people and therefore last longer (Jacobson et al., 2003).

Finally, univariate data analysis showed that the presence of the genet is positively related to the abundance of the other carnivores present in the study area considered together (*Vulpes vulpes*, *Mustela nivalis*, *Mustela putorius*, *Martes foina*, *Meles meles* and *Herepstes ichneumon*). Some of these species use the same habitats or resources, but not to the same extent or at the same time (see Fedriani et al., 1999; Mangas et al., 2005; Peris and Tena, 2005; Barrientos and Virgós, 2006). Consequently, measures related to the issues explored here might contribute not only to the conservation of the genet but also to the conservation of the whole carnivore community.

Conclusion

In this chapter, we aimed to understand the factors associated with the occurrence and abundance of several species of carnivores in agricultural landscapes of the Mediterranean region. Two main agricultural systems were studied: the open arable farmland and the montado.

Our research has shown that the less diverse carnivore assemblages occur in the largest open areas, with small remnant forests patches. These areas are dominated by generalist species as red foxes and feral dogs, which may benefit from the fragmentation induced by agricultural activities due to their high adaptability. These two species are more vagile and able to travel greater distances, exploiting new habitats, and eating different food items (e.g. fruits and/or meat) (Gittleman et al., 2001; Gehring and Swihart, 2003; Dotta and Verdade, 2007). The overabundance of these two species may limit the abundance and diversity of other smaller sized and more specialized carnivores, due to competition or even occasional predation (Garrott et al., 1993). Moreover, the enhancement of red foxes and feral dogs will increase predator pressure on key species in Mediterranean environments like the wild rabbit (*Oryctolagus cuniculus*) or the Iberian hare (*Lepus granatensis*), which are relatively abundant in Monfurado and Castro Verde, respectively (Shapira et al., 2008). The decrease in rabbit density in Iberia over the last three decades is known to be a main factor contributing to the decline of specialized carnivores of conservation concern like the Iberian lynx and the wild cat (Monterroso et al., 2009; Sarmento et al., 2005, 2009).

On the other hand, forest specialist species like genets and stone martens may face local extinction in these open areas in a short period (Hargis et al., 1999). These open areas may become impenetrable and limit the matrix permeability by hampering the easy access to food and resting sites and the dispersion of juveniles of species highly dependent upon forest areas (Lyra-Jorge et al., 2008). Consequently, the lack of sufficient number and size of permanent native forest patches will decrease the probability of persistence of these carnivores' populations (Crooks, 2002). In fact, we may find that the percentage of sample units where forest carnivores (genets and stone martens) are present is lower in the mosaic of open land and afforestation of Castro Verde than on the well conserved montado landscape in Monfurado. Probably the same trend occurs concerning carnivore abundance, although we cannot compare the abundances obtained between the two study areas once different field methodologies (transects and scent stations) were used in case studies 1 and 2 of this chapter.

The montado landscape has always been described as a habitat highly bio diverse, due to its structurally complex vegetation and traditional management system resulting in a particularly resilient and rich mosaic landscape (Ramirez and Díaz, 2008; Pita et al., 2009). In fact, the combination of holm and cork woodlands, shrublands and open grasslands interspersed by riparian galleries enhances a great animal diversity. This combination leads to a complex trophic net, which starts on soil microfauna (e.g. insects, earthworms and spiders) eaten by several insectivorous like shrews, moles, passerines, reptiles and amphibians, which in part became prey of most small and middle sized carnivores like the ones studied in this chapter. In addition, the diverse vegetation structure together with scattered grassland areas harbours great abundances of wild rabbit and red-legged partridge (Moreira et al., 2005; Delibes-Mateos et al., 2007). Consequently, hypercarnivores (meat eaters) like Iberian lynxes, wild cats and weasels (*Mustela nivalis*) and omnivorous carnivores like genets, stones martens, badgers and red foxes are able to co-exist in these landscapes.

The present favourable conservation status of both studied areas reflects the moderate human intervention that shaped these landscapes through centuries and potentiated the existence of a small and middle-sized carnivore community. However, in recent years human activities are also responsible for the main threats to these agro-systems of high conservation value and associated global biodiversity.

Economic policies promote the increase in productivity through agricultural and grazing intensification and although refrained inside Natura 2000 Sites, these changes occur in its surroundings. The intensification of agricultural practices has increased habitat fragmentation during the last years in these landscapes and in a significant part of southern Portugal (VanDoom and Bakker, 2007). Habitat fragmentation reduces landscape functional connectivity limiting wildlife movements, particularly for species with larger home ranges, like carnivores, threatening the long-term persistence of populations of these species even inside protected areas. The increase in livestock production has lead to the degradation of vegetation and watercourses while the traditional moderate levels of grazing enhanced the occurrence of some carnivores like the genet. Domestic carnivores like dogs and cats, which compete for food items with wild carnivores are increasingly overspreading due to human presence (Gittleman et al., 2001). The illegal predator control actions that can occur in some hunting estates of Southern Iberia may compromise the stability of the overall carnivore community (Duarte and Vargas, 2001; Virgós and Travaini, 2005). Recent studies in this region also highlight the negative influence roads on several wild carnivores (Galantinho and Mira, 2009; Grilo et al., 2009; Carvalho and Mira, *in press*). Most of the carnivore species are often road-killed (e.g. stone marten, genet, polecat) and some show road vicinity avoidance behaviour.

The complex relation between the crescent human activities and the community of carnivores must be understood if we wish to assure long term persistence and integrity of carnivore communities in the Mediterranean region. The continuous threat upon this biodiversity hotspot highlights our responsible conservation duty. The three studies presented in this chapter were made on Natura 2000 sites, which are extremely important areas throughout Europe. In Portugal, the proportion of Natura 2000 sites and protected areas, already reaches about 21% of the national territory (ICN, 2005). However, not all biological diversity can be preserved just inside protected areas while most of the world land is outside them (Maiorano et al., 2006; Dotta and Verdade, 2007). This is especially true for carnivores due to their large home ranges (Crooks, 2002). Enhancing connectivity measures between remnant native patches, where the links to survivorship could be attained especially by the young individuals while in dispersion seems to be fundamental for the conservation of several carnivore species. Mainly studies have to be made in human altered areas, especially in agroecosystems, once they are dominating the south Portugal landscapes (VanDoom and Bakker, 2007).

Nevertheless, to be successful we need to understand deeply the effects of human intervention and landscapes changes in the wild carnivores and find new ways to cope with change. We must also inform and involve people responsible for land use change and management about the key importance of carnivores in regulating ecosystems and the implications of land use practices in their ecology if we wish to obtain an efficient way for carnivores to exist in human altered landscapes (Riley et al., 2003).

REFERENCES

Acosta-Jamett, G. and Simonetti, J.A. (2004). Habitat use by *Oncifelis guigna* and *Pseudalopex culpaeus* in a fragmented forest landscape in central Chile. *Biodiversity and Conservation, 13*, 1135–1151.

Barea-Azcón, J.M., Virgós, E., Ballesteros-Duperón, E., Moleón, M. and Chirosa, M. (2007). Surveying carnivores at large spatial scales: a comparison of four broad-applied methods. *Biodiversity and Conservation, 16*, 1213–1230.

Barrientos, R. and Virgós, E. (2006). Reduction of potential food interference in two sympatric carnivores by sequential use of shared resources. *Acta Oecologica*, 30, 107–116.

Beja, P., Gordinho, L., Reino, L., Loureiro, F., Santos-Reis, M. and Borralho, R. (2009). Predator abundance in relation to small game management in southern Portugal: conservation implications. *European Journal of Wildlife Research, 55(3)*, 227–238.

Benton, T.G. Vickery, J.A. and Wilson, J.D. (2003). Farmland biodiversity: is habitat heterogeneity the key? *Trends in Ecology and Evolution, 18(4)*, 182–188.

Borcard, D., Legendre, P. and Drapeau, P. (1992). Partialling out the spatial component of ecological variation. *Ecology*, 73,1045–1055.

Brock, R.E. and Kelt, D.A. (2004). Influence of roads on the endangered Stephens′ Kangaroo rat (*Dipodomys stephensis*): are dirty and gravel roads different? *Biological Conservation, 118*, 633–640.

Burnham, K.P. and Anderson, D.R. (2002). *Model selection and multimodel inference: a practical information—theoretic approach* (2nd edition). New York , Springer-Verlag.

Carvalho, J.C. and Gomes, P. (2004). Feeding resource partitioning among four sympatric carnivores in the Peneda-Gerês National Park (Portugal). *Journal of Zoology, 263*, 275–283.

Carvalho, F. and Mira, A. *In press*. Comparing annual vertebrate road kills over two time periods, nine years apart: a case study in Mediterranean farmland. *European Journal of Wildlife Research*.

Conner, M.C., Labisky, R.F. and Progulske, J.R. (1983). Scent-station indices as measures of population abundance for bobcats, racoons, gray foxes and opossums. *Wildlife Society Bulletin, 11*, 146–152.

Crooks, K.R. (2002). Relative sensitivities of Mammalian Carnivores to Habitat Fragmentation. *Conservation Biology, 16(2)*, 488–502.

Cugnasse, J. and Riols, C. (1984). Contribution a la connaissance de l'écologie de la genette (*Genetta genetta* L.) dans quelques départements du sud de la France. *Gibier Faune Sauvage*, 1, 25–55.

Cuttelod, A., García, N., Abdul Malak, D., Temple, H. and Katariya, V. (2008). The Mediterranean: a biodiversity hotspot under threat. In J.-C. Vié, C. Hilton-Taylor and S.N. Stuart (Eds), *The 2008 Review of The IUCN Red List of Threatened Species* (1st edition, pp. 1–13). IUCN Gland, Switzerland.

Debano, S. (2006). Effects of livestock grazing on aboveground insect communities in semi-arid grasslands of southeastern Arizona. *Biodivers Conserv*, 15, 2547–2564.

Delibes-Mateos, M., Redpath, S.M., Angulo, E., Ferreras, P. and Villafuerte, R. (2007). Rabbits as a keystone species in southern Europe. *Biological Conservation, 137*, 149–156.

Dennis, P. (2003). Sensitivity of upland arthropod diversity to livestock grazing, vegetation structure and landform. *Journal of Food Agriculture and Environment*, 1, 2, 301–307.

Diefenbach, R.D., Conroy, J.M., Wareen, J.R., James, E.W. and Baker, A.L. (1994). A test of the Scent-Station survey Technique for bobcats, *Journal of Wildlife Management, 58(1)*, 10–17.

Disney, M.R., Hellgren, E.C., Davis, G.A., Leslie, Jr., D.M. and Engle, D.M. (2008). Relative abundance of mesopredators and size of oak patches in the cross timbers ecoregion. *The Southwestern Naturalist, 53(2)*, 214–223.

Dobkin, D.S., Rich, A.C. and Pyle, W.H. (1998). Habitat and avifaunal recovery from livestock grazing in a riparian meadow system of the Northwestern Great Basin. *Conservation Biology*, 12, 1, 209–221.

Dotta, G. and Verdade, L.M. (2007). Trophic categories in mammal assemblage: diversity in agricultural landscape. *Biota Neotropica, 7(2)*, 287-292.

Duarte, J. and Vargas, J.M. (2001). Son selectivos los controles de predadores en los cotos de caza? *Galemys*, 13, special number, 1–9.

Eikie, P.C., Rempel, R.S. and Carr, A.P. (1999). *Patch analyst user's manual. A tool for quantifying landscape structure.* NWST Technical Manual TM–002.

ERENA. (2004). *Inventário florestal e cartografia de ocupação do solo do sítio da Serra de Monfurado.* Ordenamento e Gestão de Recursos Naturais, Lda.

Espírito-Santo, C., Rosalino, L.M. and Santos-Reis, M. (2007). Factors affecting the placement of common genet latrine sites in a Mediterranean Landscape in Portugal. *Journal of mammalogy, 88*, 201–207.

ESRI. (1999). *ArcView GIS 3.2 ESRI.* Environmental Systems Research Institute, CA, USA Inc. New York.

Fedriani, J.M., Palomares, F. and Delibes, M. (1999). Niche relations among three sympatric Mediterranean carnivores. *Oecologia*, 121, 138–148.

Fleischner, T.L. (1994). Ecological costs of livestock grazing in Western North America. *Conservation Biology*, 8, 629–644.

Galantinho, A. and Mira, A. (2009). The influence of human, livestock, and ecological features on the occurrence of genet (*Genetta genetta*): a case study on Mediterranean farmland. *Ecological Research, 24(3)*, 671–685.

Garrott, R.A., White, P.J. and Vanderblit White, C.A. (1993). Overabundance: an issue for conservation biologists? *Conservation Biology, 7*, 946–949.

Gehring, T.M. and Swihart, R.K. (2003). Body size, niche breath, and ecologically scaled responses to habitat fragmentation: mammalian predators in an agricultural landscape. *Biological Conservation, 109*, 283–295.

Ginsberg, J.R. (2001). Setting priorities for carnivore conservation: what makes carnivores different? In J.L. Gittleman, S.M. Funk, D. Macdonald and R.K. Wayne (Eds.), *Carnivore Conservation: Conservation Biology 5* (1st edition, pp. 498–523). Cambridge: The Zoological Society of London, Cambridge University Press.

GIEWS-FAO (2009). *Crops prospects and food situation.* No. 3. Available at http://www.fao.org/docrep/012/ai484e/ai484e00.htm

Gittleman, J.L., Funk, S.M., Macdonald, D. and Wayne, R.K. (2001). *Carnivore Conservation: Conservation Biology 5.* (1st edition).

Cambridge: The Zoological Society of London, Cambridge University Press.

Gomes, P. and Giraudoux, P. (1992). Structure du paysage et habitat de la Genette (*G. genetta*, L.). *XV Colloque Francophone de Mammalogie: Les Carnivores*, nº spécial, 59–63.

Gompper, M.E., Kays, R.W., Ray, J.C., LaPoint, S.D., Bogan, D.A. and Cryan, R.J. (2006). A comparison of noninvasive techniques to survey carnivore communities in Northeastern North America. *Wildlife Society Bulletin, 34*, 1142–1151.

Grilo, C., Bissonette, J. and Santos–Reis, M. (2009). Spatial-temporal patterns in Mediterranean carnivore road casualties: Consequences for mitigation. *Biological Conservation, 142*, 301–313.

Groom, M.J. and Vynne, C.H. (2006). Habitat degradation and loss. In M.J. Groom, G.K. Meffe and C.R. Carroll (Eds.), *Principles of Conservation Biology* (3rd edition, pp. 173–212). Sunderland, Massachusetts, U.S.A.: Sinauer Associates Inc.

Hargis, D.C., Bissonette, A.J. and Turner, L.D. (1999). The influence of forest fragmentation and landscape pattern on American martens. *Journal of Applied Ecology, 36*, 157–172.

Henle, K., Alard, D., Clitherow, J., Cobb, P., Firbank, L., Kull, T., McCracken, D., Moritz, R.F.A., Niemelä, J. Rebane, M. Wascher, D., Watt, A and Young, J. (2008). Identifying and managing the conflicts between agriculture and biodiversity conservation in Europe-A review. *Agriculture, Ecosystems and Environment, 124*, 60–71.

Hilty J.A., Brooks, C., Heaton, E. and Merenlender, A.M. (2006). Forecasting the effect of land-use change on native and non-native mammalian predator distributions. *Biodiversity and Conservation, 15*, 2853–2871.

Hilty J.A. and Merenlender A.M. (2004). Use of riparian corridors and vineyards by mammalian predators in Northern California. *Conservation Biology, 18*, 126–135.

Hosmer, J.R. and Lemeshow, D.W. (2000). *Applied Logistic Regression.* (2nd edition) New York: John Wiley and Sons.

ICN (2005). Plano Sectorial da Rede Natura 2000. Relatório Preliminar, vol I. Instituto da Conservaçãoo da Natureza (ICN), Lisboa.

IUCN (2007). 2007 IUCN Red List of Threatened Species. Available at http://www.iucnredlist.org

Jacobson, S., Sieving, K., Jones, G. and Van Doorn, A. (2003). Assessment of framer attitudes and behavioural intentions toward bird conservation on organic and conventional Florida farms. *Conservation Biology, 17(2),* 595–606.

Joffre, R., Rambal, S. and Ratte, J.P. (1999). The dehesa system of southern Spain and Portugal as a natural ecosystem mimic. *Agroforestry Systems, 45,* 57–79.

Krueper, D., Bart, J., Rich, T.D. (2003). Response of vegetation and breeding birds to the removal of cattle on the San Pedro River, Arizona USA. *Conservation Biology*, 17,2, 607–615.

Lee, J. and Wong, D. (2001). *Statistical Analysis with Arcview GIS®*. (1st edition). John Wiley and Sons, Inc.

Livaitis, J.A. (2001). Importance of early successional habitats to mammals in eastern forests. *Wildlife Society Bulletin, 29(2)*, 466–473.

Long, R.A., MacKay, P., Zielinski, W.J. and Ray, J.C. (2008). *Noninvasive Survey Methods for Carnivores.* (1st edition). Island Press, Washington, DC.

Lyra-Jorge, M.C., Ciocheti, G. and Pivello, V.R. (2008). Carnivore mammals in a fragmented landscape in northeast of São Paulo State, Brazil. *Biodiversity and Conservation, 17*, 1573–1580.

Lyra-Jorge, M.C., Ribeiro, M.C., Clocheti, G., Tambosi, L.R. and Pivello, V.R. (2010). Influence of multiscale landscape structure on the occurence of carnivorous mammals in a human-modified savanna, Brazil. *European Journal of Wildlife Research*

Macdonald, D. (2004). *The new encyclopedia of mammals.* (1st edition). Oxford University Press, Oxford.

Macdonald, D.W. and Barrett, P. (1993). *Mammals of Britain and Europe.* (1st edition). London: Harper Collins.

Macdonald, D.W. and Rushton, S. (2003). Modelling space use and dispersal of mammals in real landscapes: a tool for conservation. *Journal of Biogeography, 30,* 607–620.

Maiorano, L., Falcucci, A. and Boitani, L. (2006). Gap analysis of terrestrial vertebrates in Italy: Priorities for conservation planning in human dominated landscape. *Biological Conservation, 133*, 455–473.

Manel, S., Williams, H.C. and Ormerod, S.J. (2001). Evaluating presence-absence models in ecology: the need to account for prevalence. *Journal of Applied Ecology*, 38, 921–931.

Mangas, J.G., Carrobles, M., Alcázar, L.H., Bellón, D. and Virgós, E. (2005). Estudio preliminar de la distribución de la garduña (*Martes martes*) y la gineta (*Genetta genetta*) en un pinar mediterráneo del centro de la

Península Ibérica. Abstracts of VII Jornadas de la Sociedad Española para la Conservación y Estudio de los Mamíferos, SECEM, Valência, pp. 121.

Mangas, K.G., Lozano, J., Cabezas-Díaz, S. and Virgós, E. (2008). The priority value of scrubland habitats for carnivore conservation in Mediterranean ecosystems. *Biodiversity and Conservation, 17,* 43–51.

Marrs, R.H., Galtress, K., Tong, C., Cox, E.S., Blackbird, S.J., Heyes, T.J., Pakeman, R.J. and Le Duc, M.G. (2007). Competing conservation goals, biodiversity or ecosystem services: element losses and species recruitment in a managed moorland: bracken model system. *Journal of Environmental Management,* 85,4, 1034–1047.

Matos, H.M., Santos, M.J., Palomares, F. and Santos-Reis, M. (2009). Does riparian habitat condition influence mammalian carnivore abundance in Mediterranean ecosystems? *Biodiversity and Conservation, 18*, 373–386.

Mazerolle, J.M. and Villard, A.M. (1999). Patch characteristics and landscape context as predictors of species presence and abundance: A review. *Ecoscience, 6 (1)*, 117–124.

Michalski, F. and Peres, C.A. (2005). Anthropogenic determinants of primate and carnivore local extinctions in a fragmented forest landscape of southern Amazonia. *Biological Conservation, 124,* 383–396.

Mickevicius, E. (2002). Distribution of badger (*Meles meles*), fox (*Vulpes vulpes*) and raccoon dog (*Nyctereutes procyonoides*) burrows in different habitats and soil types of Lithuania. *Acta Zoologica Lithuanica, 12*, 159–166.

Monterroso, P., Brito, J.C., Ferreras, P. and Alves, P.C. (2009). Spatial ecology of the European wildcat in Mediterranean ecosystem: dealing with small radio-tracking datasets in species conservation. *Journal of Zoology*, 279, 27–35.

Moore, N., Whiterow, A., Kelly, P., Garthwaite, D., Bishop, J., Langton, S. and Cheeseman, C. (1999). Survey of badger *Meles meles* damage to agriculture in England and Wales. *Journal of Applied Ecology, 36*, 974–988.

Moreira, F., Beja, P., Morgado, R., Reino, L., Gordinho, L., Delgado, A. and Borralho, R. (2005). Effects of field management and landscape context on grassland wintering birds in Southern Portugal. *Agriculture, Ecosystem and Environment, 109,* 59–74.

Moreno, G. (2008). Response of understorey forage to multiple tree effects in Iberian dehesas. *Agriculture, Ecosystems and Environment*, 123, 239–244.

Moreno, G., Obrador, J.J. and García, A. (2007). Impact of evergreen oaks on soil fertility and crop production in intercropped dehesas. *Agriculture, Ecosystems and Environment*, 119, 270–280.

Myers, N., Mittermeier, R.A., Mittermeier, C.G., da Fonseca, G.A.B. and Kent, J. (2000). Biodiversity hotspots for conservation priorities. *Nature, 403,* 853–858.

Nugent, L.K., Sihanonth, P., Thienhirun, S. and Whalley, A.J.S. (2005). Biscogniauxia: a genus of latent invaders. *Mycologist, 19(1)*, 40–43.

OECD-FAO (2009). Agricultural Outlook 2009-2018: Highlights. OECD Publishing.

Oehler, J.D. and Livatis, J.A. (1996). The role of spatial scale in understanding responses by medium-sized carnivores to forest fragmentation. *Canadian Journal of Zoology, 74,* 2070–2079.

Olden, J.D., Jackson, D.A. and Peres-Neto, P.R. (2002). Predictive models of fish species distributions: a note on proper validation and chance predictions. *Transactions of the American Fisheries Society, 131*, 329-336.

Palomares, F. (1993). Faecal marking behaviour by free-ranging Common genets *Genetta genetta* and Egyptian mongooses *Herpestes ichneumon* in southwestern Spain. *Z Saugetierkd*, 58, 225–231.

Palomares, F. and Caro, T.M. (1999). Interspecific killing among mammalian carnivores. *The American Naturalist, 153(5)*, 492–508.

Palomares, F. and Delibes, M. (1994). Spatio-Temporal Ecology and behaviour of European Genets in south-western Spain. *Journal of Mammalogy, 75 (3)*, 714–724.

Palomares, F. and Delibes, M. (1988). Time and space use by two common genets (*Genetta genetta*) in the Doñana national park, Spain. *Journal of Mammology, 69 (3)*, 635–637.

Peris, A. and Tena, L. (2005). Distribuición y selección de habitat de la gineta (*Genetta genetta*) e de la garduña (*Martes foina*) en un ecosistema mediterráneo. *Abstracts of VII Jornadas de la Sociedad Española para la Conservación y Estudio de los Mamíferos*, SECEM, Valência, p 120.

Pinto-Correia, T. (1993). Threatened landscape in Alentejo, Portugal: the montado and other agro-silvo-pastoral systems. *Landscape and Urban Planning 24*, 43–48.

Pinto-Correia, T. and Vos, W. (2004). Multifunctionality in Mediterranean landscapes – past and future. In R.H.G. Jongman (Ed.) *The new dimensions of the European landscape* (1st edition, pp. 135–164). Dordrecht: Springer.

Pita, R., Mira, A., Moreira, F., Morgado, R. and Beja, P. (2009). Influence of landscape characteristics on carnivore diversity and abundance in Mediterranean farmland. *Agriculture, Ecosystems and Environment, 132,* 57–65.

Queirós, A.I. (1989). Regime alimentar de *Genetta genetta* L. e sua relação trófica no interior da comunidade de mamíferos. Serviço Nacional de Parques, Reservas e Conservação da Natureza, Lisboa.

Ramirez, J.A. and Díaz, M. (2008). The role of temporal shrub encroachment for the maintenance of Spanish holm oak *Quercus ilex* dehesas. *Forest Ecology and Management, 255*, 1976–1983.

Rawlings, J.O. (1988). *Applied regression analysis: a research tool.* Wadsworth, Inc., Belmont.

Reino, L., Beja P., Osborne, P.E., Morgado, R., Fabião, A. and Rotenberry, J.T. (2009). Distance to edges, edge contrast and landscape fragmentation: interactions affecting farmland birds around forest plantations. *Biological Conservation, 142,* 824–838.

Revilla, E., Palomares, F. and Delibes, M. (2000). Defining key habitats for low density populations of Eurasian badgers in Mediterranean environments. *Biological Conservation, 95*, 269–277.

Revilla, E., Palomares, F. and Fernández, N. (2001). Characteristics, location and selection of diurnal resting dens by Eurasian badgers (Meles meles) in a low density area. *Journal of Zoology, 255*, 291-299.

Riley, S.P.D., Sauvajot, R.M., Fuller, T.K., York, E.C., Kamradt, D.A., Bromley, C. and Wayne, R.K. (2003). Effects of urbanization and habitat fragmentation on bobcats and coyotes in southern California. *Conservation Biology, 17(2)*, 566–576.

Rivas-Martinez, S. and Loidi, J. (1999). Biogeography of the Iberian Peninsula. *Itinera Geobotanica*, 13, 49–67.

Roeder, J-J. (1980). Les emplacements de defecation chez la Genette *Genetta genetta* L. *Revue d'Ecologie (Terre and Vie)*, 34, 485–494.

Rondinini, C. and Boitani, L. (2002). Habitat use by beech martens in a fragmented landscape. *Ecography, 25*, 257–264.

Rosalino, L.M., Loureiro, F., MacDonald, D.W. and Santos-Reis, M. (2003). Food digestibility of a Eurasian badger *Meles meles* with special reference to the Mediterranean region. *Acta Theriologica*, 48, 283–288.

Rosalino, L.M., Loureiro, F., MacDonald, D.W. and Santos-Reis, M. (2005b). Dietary shifts of the badger *Meles meles* in Mediterranean woodlands: an opportunistic forager with seasonal specialisms. *Mammalian. Biology, 70*, 12–23.

Rosalino L.M., Macdonald D.W. and Santos-Reis M. (2005a). Activity rhythms, movements and patterns of sett use by badgers, *Meles meles*, in a Mediterranean woodland. *Mammalia, 69(3-4)*, 395–408.

Rosalino, L.M., Rosário, J. and Santos-Reis, M. (2009). The role of habitat patches on mammalian diversity in cork oak agroforestry systems. *Acta Oecologica, 35,* 507–512.

Rosalino, L.M. and Santos-Reis, M. (2002). Feeding habits of the common genet *Genetta genetta* in a man-made landscape of central Portugal. *Mammalia, 66(2),* 195–205.

Santos, M.J., Pedroso, N.M., Ferreira, J.P., Matos, H.M., Sales-Luís, T., Pereira, I., Baltazar, C., Grilo, C., Cândido, A.T., Sousa, I. and Santos-Reis, M.. (2008). Assessing dam implementation impact on threatened carnivores: the case of Alqueva in SE Portugal. *Environment and Monitoring Assessment, 142*, 47–64.

Santos-Reis, M. and Mathias, M.L. (1998). The historical and recent distribution and status of mammals in Portugal. *Hystrix*, 8, 75–89.

Santos-Reis, M., Santos, M.J., Lourenço, S., Marques, J.T., Pereira, I. and Pinto, B. (2004). Relationships between stone martens, genets and cork oak woodlands in Portugal. In D. J. Harrison, A. K. Fuller and G. Proulx (Eds.), *Marten and fishers (Martes) in human-altered environments: an international perspective* (1st edition, pp. 147–172). New York: Springer Science and Business Media Inc.

Sargeant, G.A., Johnson, D.H. and Berg, W.E. (2003). Sampling designs for carnivore scent-station surveys. *Journal of Wildlife Management, 67(2)*, 289–298.

Sarmento, P., Cruz, J., Monterroso, P., Tarroso, P., Ferreira, C. and Negrões, N. (2005). Iberian Lynx conservation in Portugal: dilemmas and solutions. Wildlife Biology and Practice, 1(2), 156–162.

Sarmento, P., Cruz, J., Monterroso, P., Tarroso, P., Ferreira, C. and Negrões, N. (2009). Status survey of the critically endangered Iberian Lynx *Lynx pardinus* in Portugal. *European Journal of Wildlife Research, 55*, 247–253.

Shapira, I., Sultan, H. and Shanas, U. (2008). Agricultural farming alters predator–prey interactions in nearby natural habitats. *Animal Conservation, 11,* 1–8.

Sokal, R.R. and Rohlf, J.R. (1997). *Biometry: the principles and practice of statistic in biological research.* (3st editon). W. H. Freeman and Company. New York.

Stephens, D.W. and Krebs, J.R. (1986). *Foraging theory*. New Jersey, Princeton University Press.

Sunquist, M.E. and Sunquist, F. (2001). Changing landscapes: consequences for carnivores. In J.L. Gittleman, S.M. Funk, D. Macdonald and R.K. Wayne (Eds.), *Carnivore conservation: Conservation biology 5* (1st edition, pp. 399–418). Cambridge: The Zoological Society of London, Cambridge University Press.

TabachnicK, B.G. and Fidell, L.S. (2001). *Using Multivariate Statistics*. (4th edition). Allyn and Bacon, Boston.

VanDoom, A.M. and Bakker, M.M. (2007). The destination of arable land in a marginal agricultural landscape in South Portugal: an exploration of land use change determinants. *Landscape Ecology, 22*, 1073–1087.

VanDoorn, A.M. and Pinto-Correira, T. (2007). Differences in land cover interpretations in landscapes rich in cover gradients: reflections based on *Montado* of South Portugal. *Agroforestry Systems, 70*, 169–183.

Virgós, E. (2001). Relative value of riparian woodlands in landscapes with different forest cover for medium-sized Iberian carnivores. *Biodiversity and Conservation, 10,* 1039–1049.

Virgós, E. and Casanovas, J.G. (1997). Habitat selection of *Genetta genetta* in the mountains of central Spain. *Acta Theriologica, 42 (2),* 169–177.

Virgós, E. and Garcia, F.J. (2002). Patch occupancy by stone martens *Martes foina* in fragmented landscapes of central Spain: the role of fragment size, isolation and habitat structure. *Acta Oecologica, 23*, 231–237.

Virgós, E., Lorente, M. and Cortés, Y. (1999). Geographical variation in genet (*Genetta genetta* L.) diet: a literature review. *Mammal Review, 29*, 117–126.

Virgós, E., Romero, T. and Mangas, J.G. (2001). Factors determining ''gaps'' in the distribution of a small carnivore, the common genet (*Genetta genetta*), in central Spain. *Canadian Journal of Zoology,* 79, 1544–1551.

Virgós, E., Tellería, J.L. and Santos, T. (2002). A comparison on the response to forest fragmentation by medium-sized Iberian carnivores in central Spain. *Biodiversity and Conservation, 11,* 1063–1079.

Virgós, E. and Travaini, A. (2005). Relationship between small-game hunting and carnivore diversity in Central Spain. *Biodiversity and Conservation, 14*, 3475–3486.

Wade, M.R., Gurr, G.M. and Wratten, S.D. (2008). Ecological restoration of farmland: progress and prospects. *Philosophical Transactions of the Royal Society B, 363,* 831–847.

Wheeler, P. (2008). Effects of sheep grazing on abundance and predators of field vole (Microtus agrestis) in upland Britain. *Agriculture Ecosystems and Environment*, 123, 49–55.

Wilson, G.J. and Delahay, R.J. (2001). A review of methods to estimate the abundance of terrestrial carnivores using field signs and observation. *Wildlife Research, 28*,151–164.

Wolters, V. (2000). Invertebrate control of soil organic matter stability. *Biology and Fertility of Soils* 31, 1–19.

Woodcock, B.A., Pywell, R.F., Roy, D.B., Rose, R.J. and Bell, D. (2005). Grazing management of calcareous grasslands and its implications for the conservation of beetle communities. *Biological Conservation*, 125, 193–202.

Yanes, M. and Suarez, F. (1995). Nest predation patterns in ground-nesting passerines on the Iberian Peninsula. *Ecography, 18*, 423–428.

Zabala, J., Zuberogoitia, I., Garin, I. and Aihartza, J. (2003). Landscape features in the habitat selection of European mink (*Mustela lutreola*) in south-western Europe. *Journal of Zoology, 260*, 415–421.

Zielinski, W.J. (2000). Weasels and martens – carnivores in northern latitudes. In Halle, S. and Stenseth, N.C. (eds.) *Activity patterns in small mammals.* (1st edition, pp. 95-118). Ecological Studies, Vol. 141. Springer-Verlag, Berlin Heidelberg.

Zuur, A., Ieno, E. and Graham, S. (2007). *Analysing ecological data.* (1st edition). Springer, UK.

INDEX

A

B

C

F

G

H

I

J

K

L

M

N

O

P

T

U

V

W

X

Y

Z